# Studies in Computational Intelligence

Volume 724

**Series editor**

Janusz Kacprzyk, Polish Academy of Sciences, Warsaw, Poland
e-mail: kacprzyk@ibspan.waw.pl

*About this Series*

The series "Studies in Computational Intelligence" (SCI) publishes new developments and advances in the various areas of computational intelligence—quickly and
with a high quality. The intent is to cover the theory, applications, and design
methods of computational intelligence, as embedded in the fields of engineering,
computer science, physics and life sciences, as well as the methodologies behind
them. The series contains monographs, lecture notes and edited volumes in
computational intelligence spanning the areas of neural networks, connectionist
systems, genetic algorithms, evolutionary computation, artificial intelligence,
cellular automata, self-organizing systems, soft computing, fuzzy systems, and
hybrid intelligent systems. Of particular value to both the contributors and the
readership are the short publication timeframe and the worldwide distribution,
which enable both wide and rapid dissemination of research output.

More information about this series at http://www.springer.com/series/7092

Amit Konar · Sriparna Saha

# Gesture Recognition

## Principles, Techniques and Applications

 Springer

Amit Konar
Artificial Intelligence Laboratory,
    Department of Electronics and
    Tele-Communication Engineering
Jadavpur University
Kolkata
India

Sriparna Saha
Artificial Intelligence Laboratory,
    Department of Electronics and
    Tele-Communication Engineering
Jadavpur University
Kolkata
India

ISSN 1860-949X          ISSN 1860-9503   (electronic)
Studies in Computational Intelligence
ISBN 978-3-319-87259-9          ISBN 978-3-319-62212-5   (eBook)
DOI 10.1007/978-3-319-62212-5

Printed on acid-free paper

This Springer imprint is published by Springer Nature
The registered company is Springer International Publishing AG
The registered company address is: Gewerbestrasse 11, 6330 Cham, Switzerland

# Preface

The book to the best of the authors' knowledge is the first comprehensive title of its kind that provides a clear and insightful introduction to the subject in a precise and easy-to-read manner. It examines gesture recognition as a pattern recognition problem, and proposes intelligent techniques to efficiently solve problems of diverse domains, covering dance, healthcare and gaming.

Gesture is non-vocal communication, synthesized with the help of noticeable body movements, conveying certain messages. Humans generally involve their arms, legs and voluntary muscles of their faces and other organs to represent their gesture. Each gesture thus has a body language, which usually is unique within a group/society and is sometimes induced by culture and inheritance. Gesture recognition is important to improve interactions between the humans and the computer. Automated recognition of human gesture has wide-spread applications, including healthcare, e-learning of dances, emotion understanding of psychological patients.

Like most of the common pattern recognition problems, gesture recognition can be done by unimodal as well as multi-modal means. The unimodal means require more precise features for accurate classification of gestural data into two or more classes. Multi-modality has an advantage that individual modalities express diversity of the gestural information from different standpoints, and thus often is found to be synergistic rather than competitive. The book, however, primarily focuses on unimodal gestural data, such as pixel intensity when gestural information is captured in images or video clips. Alternatively, the book deals with gestural data obtained from Microsoft Kinect, where a frame of sampled Kinect data contains 20 junction coordinates of the subject. These data are pre-processed to keep them free from measurement noise that appears in the raw image/Kinect data because of environmental noise/noisy ambience. The book also covers one exemplary chapter on multi-modal data, where electro-encephalographic modality is used to acquire the brain states of the experimental subject and Kinect data to acquire body-junction coordinates.

Gesture recognition being a pattern classification problem employs a supervised learning technique, where a set of training instances obtained from real raw gestural data and its user-defined class are used to build up the training instances. Any (traditional or new) supervised learning algorithm may be employed to train the mapping between the input and output training instances. After the mapping function is built up by a suitable machine learning algorithm, it can be utilized to generate the response for an unknown input stimulus. The choice of the supervised learning algorithm depends on execution time of the classification algorithm. A fast algorithm always gets a higher preference. However, the sampling rate of the raw data truly determines the classifier algorithm as most of the data processing is performed within one sampling interval to keep the gesture recognition system amenable for real-time application. The well-known off-the-shelf algorithms of pattern classification that can serve the purpose includes Back-propagation algorithm, Support Vector Machine (SVM) and Linear/Quadratic Discriminant Analysis (LDA/QDA) techniques. Although Naïve Bayes' is well-acclaimed in pattern classification, it requires quite a large number of conditional probabilities, and thus is time-costly for realization.

Gesture recognition is hard when the sensory readings recorded are not free from measurement noise. Such noise in dance video/images may appear because of poor ambience and lighting condition. Non-imaging recording systems, such as Microsoft Kinect, may also add noise to the measurements because of poor penetration of the infrared due to infrared absorbing garments. The other possibility of losing accurate gestural data by Kinect is due to out of sensor ranges, presence of fluorescent light in the vicinity of the Kinect. Besides the above, there exist intra-personal and inter-personal variations in gestural data. For example, a person may have several gestures, when suffering from joint pains at her knee. Further, the recognition problem becomes harder when the subjects have their variation for the same gestural information, such as having similar gesture while having a similar pain at knee or specific location of her other joints. The book offers a generic solution to the above problem by incorporating type-2 fuzzy sets.

An Interval Type-2 Fuzzy Set (IT2FS) usually is expressed by two type-1 membership functions, called Upper and Lower Membership Functions that together fixes the span of uncertainty, lying within the said bounds. The wider the containment (also called the footprint of uncertainty), the larger is the uncertainty involved in the measurement process. Besides IT2FS, there exist a more generic version of type-2 fuzzy sets, called General Type-2 Fuzzy Sets (GT2FS), currently gaining increasing popularity in the type-2 literature. GT2FS has more flexibility of representing intra- and inter-personal level uncertainty than the IT2FS counterpart. However, because of its excessive computational overhead it is rarely used in type-2 literature. In this book, we however, deal with both IT2FS- and GT2FS-induced classifiers to recognize gestures captured from sources, which are not totally free from imprecision and uncertainty. A novel approach for IT2FS and GT2FS induced gesture classification is proposed to classify ambiguous Kinect features that might fall in more than one classes.

The book includes nine chapters covering various aspects of gesture recognition related to day-to-day gestures and also some dance forms with special reference to different computational intelligence techniques used for this purpose.

Chapter 1 outlines basic concepts on gesture recognition. It begins with foundations in gesture recognition. Next, we introduce Microsoft Kinect as a gesture acquisition device. Later part of the chapter offers a thorough review on gesture recognition, covering both whole body and hand gestures. The chapter finally summarizes the scope and possible application areas of gesture recognition.

Chapter 2 targets online recognition of ballet primitives for e-learning applications. For this purpose, novice dancers are asked to enact the given ballet primitives in front the camera. Euler number and Radon transform are taken as the features from each ballet posture and recognition of unknown posture is achieved by correlation.

Chapter 3 provides another interesting approach to posture recognition in ballet dance using fuzzy logic. Whenever large numbers of dancers are asked to perform a specific ballet posture, then a large amount of uncertainty is noticed among the ballet postures. To tackle this problem, fuzzy logic is introduced in this chapter.

The lately developed Microsoft's Kinect sensor has opened an innovative horizon to recognize human body using three-dimensional body joints. In Chap. 4, a novel gesture driven car racing game control is proposed using Mamdani type-I fuzzy inference for providing better experience of the game. Instead of pressing keys in the keyboard, here gamers can get the feeling of rotating a virtual steering wheel by moving their hands accordingly. The acceleration and brake (deceleration) of the vehicle is controlled by the relative displacement of the right and the left foot.

In today's' era, most of the families are nuclear and many elderly people live alone in their homes. Home monitoring systems are gaining much importance in this socio-economic condition. The gestural features of a subject suffering from the same pathological disorder vary widely over different instances. Wider variance in gestural features is the main source of uncertainty in the pathological disorder recognition, which has been addressed in Chap. 5 using type-2 fuzzy sets.

For providing better human computer interaction for e-learning of different dances, Kinect sensor is introduced for human gesture detection in Chap. 6. The elementary knowledge of geometry has been employed to introduce the concept of planes in the feature extraction stage. The gestures are classified using a probabilistic neural network.

In Chap. 7, a unique idea of dance composition is proposed. This system automatically composes dance sequences from previously stored dance movements by implementing differential evolution algorithm. Gestural data, specific of dance moves, is captured using a Microsoft Kinect sensor. An algorithm is developed to fuse various Indian classical dances of same theme and rhythm.

A highly essential application for rehabilitative patients is provided in Chap. 8. The chapter focuses on directing an artificial limb based on brain signals and body gestures to assist the rehabilitative patients. To accomplish the timing constraint parameter of real-time system, cascade correlation learning architecture is implemented here.

Finally Chap. 9 concludes the book with discussion about all previous chapters.

Kolkata, India                                                                  Amit Konar
April 2017                                                                     Sriparna Saha

# Acknowledgements

Authors gratefully acknowledge the support they received from individuals to write the book in its current form. Although the authors cite the contributions of a few individuals, they are grateful to all people who have helped in whatever way possible for the development of the book.

The first author gratefully acknowledges the support he received from his colleagues, friends and students, without whose help and support the book could not be prepared in its current form. The second author equally is grateful to all his colleagues of Maulana Abul Kalam Azad University of Technology (formerly known as West Bengal University of Technology), Kolkata, whose unfailing help and assistance nurtured the progress of the book. The authors would also like to thank Prof. Suranjan Das, Vice-Chancellor of Jadavpur University and Prof. Saikat Maitra, Vice Chancellor, MAKAUT for providing them good and resourceful environments to develop the book. The authors are equally indebted to Prof. P. Venkateswaran, HOD, Department of Electronics and Telecommunication Engineering, Jadavpur University for providing all sorts of support the authors required to complete the book in the present form. Authors are grateful to the entire Springer for their hard work and kind support at different levels of production of the book.

Lastly, the authors acknowledge the support they received from their family members. In this regard, the first author recognizes the support of his wife: Srilekha, son: Dipanjan and mother: Mrs. Minati Konar. The second author also acknowledges the support she received from her parents: Dr. Sukla Saha and Dr. Subhasish Saha.

April 2017
Amit Konar
Sriparna Saha

# Contents

# About the Authors

**Amit Konar** (SM'10) received the B.E. degree from Indian Institute of Engineering Science and Technology (formerly Bengal Engineering College), Howrah, India, in 1983, and the M.E., M.Phil., and Ph.D. degrees from Jadavpur University, Kolkata, India, in 1985, 1988, and 1994, respectively.

He is currently Professor with the Department of Electronics and Telecommunication Engineering, Jadavpur University, where he is also the Founding Coordinator of the M.Tech. Program on Intelligent Automation and Robotics.

He has supervised 21 Ph.D. theses. He has over 300 publications in international journals and conference proceedings. He is the author of eight books, including two popular textbooks, Artificial Intelligence and Soft Computing (Boca Raton, FL, USA: CRC Press, 2000) and Computational Intelligence: Principles, Techniques and Applications (New York, NY, USA: Springer, 2005).

Dr. Konar is currently serving as an Associate Editor of IEEE Trans. on Fuzzy Systems, IEEE Trans. on Emerging Topics in Computational Intelligence and Neurocomputing, Elsevier. He also served as an Associate Editor of IEEE Trans. on Systems, Man and Cybernetics: Part-A and Applied Soft Computing, Elsevier. He is the Editor-in-Chief of a Springer Book Series on Cognitive Intelligence and Robotics.

**Sriparna Saha** received her M.E. and Ph.D. degrees from Electronics and Telecommunication Engineering Department, Jadavpur University. She has published extensively on image analysis and gesture recognition. She had offered post-graduate level courses in Intelligent Automation and Robotics, Jadavpur University, India. She is currently an assistant professor of Maulana Abul Kalam Azad University of Technology (formerly known as WBUT), Kolkata, India.

# Chapter 1
# Introduction

The chapter provides an introduction to gesture recognition. It begins with a thorough review of the principles of gesture recognition undertaken in the existing works. The chapter then elaborately introduces the Kinect sensor, which has recently emerged as an important machinery to capture human-gestures. Various tools and techniques relevant to image processing, pattern recognition and computational intelligence, which have necessary applications in gesture recognition, are also briefly explained here. The chapter outlines possible applications of gesture recognition. The scope of the book is also appended at the end of the chapter.

## 1.1 Defining Gestures

The word "gesture" refers to orientation of the whole/part of human physiques to convey non-verbal information, representing emotions, victory in games/war, traffic control/signalling by police and many others. Although gestures and postures are synonymous, gestures carry more information than postures as it includes facial expression as well with body language.

According to Darwin [1], individual species, for instance cats and dogs, carry universal gestures irrespective of their geographical locations of their habitats and cultural environment in which they are brought up. However, humans of different geographical locations and/or environment sometimes employ different gestures to convey the same non-verbal information. For example, conveying regards to seniors in people of different cultures/societies are different.

Gestures include both static and dynamic description of human physiques [2]. A static gesture, as evident from its name, is a fixed orientation of the components/fragments of our external organs that together conveys the non-verbal message to others. Examples of static gesture include a victory sign (represented by a V formed by the forefinger and the middle finger of our palm), finger-pointing to an object/person for referencing, and putting palms together to indicate gratitude to

© Springer International Publishing AG 2018
A. Konar and S. Saha, *Gesture Recognition*, Studies in Computational
Intelligence 724, DOI 10.1007/978-3-319-62212-5_1

God or respectable elderly people. Dynamic gestures include motion of one's body part, as used in signalling by police or by referees in certain outdoor tournaments. Besides static and dynamic gestures, we have mixed gestures, containing partly static and partly dynamic. Dances, for instance, include both static and dynamic gestures to represent complex emotions, theme and culture.

Irrespective of static and dynamic characteristics of gestures, we can classify them into two major heads: Intrinsic and Extrinsic gestures (Fig. 1.2). Intrinsic gestures refer to latent subjective intent and meant for oneself and not for communication to others for better explanation of any situation and/or theme. Examples of intrinsic gestures include subjective look, while pondering over a solution of a complex problem/situation (Fig. 1.1a), anxious thought, represented by finger biting (Fig. 1.1b) and shying (Fig. 1.1c). Extrinsic gestures are meant for other people to communicate certain messages. We classify extrinsic gestures into two major classes: Emotional gestures and Non-emotional gestures. Emotional gestures again could be of two types, gestures conveying positive emotions and negative emotions. Figure 1.3 provides different emotional gestures. Non-emotional gesture again is classified into several types. Instructing gestures, signalling gestures and responsive gestures are just a few falling under non-emotional gestures. Instructive gestures refer to gestures conveying certain instructions, such as "come here" or "go there" (Fig. 1.4a, b). It is noteworthy that "come here" is a dynamic gesture and so represented by consecutive palm orientation over time. Signalling gestures were earlier used by traffic police and still prevalent in a few countries. Driving gestures of vehicle drivers also carry certain signalling message to pedestrians or other drivers. Driving gestures too fall under signalling gestures. Figure 1.5 provides illustrative driving gestures. Responsive gestures, such as supporting or refuting a speaker by a listener, is also an example of non-emotional gestures. Figure 1.2 presents the gesture classification tree. This, of course, is illustrative and may be expanded or altered and thus is not unique.

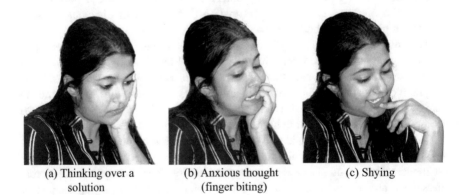

(a) Thinking over a          (b) Anxious thought          (c) Shying
    solution                     (finger biting)

**Fig. 1.1** Pictorial view of intrinsic gestures

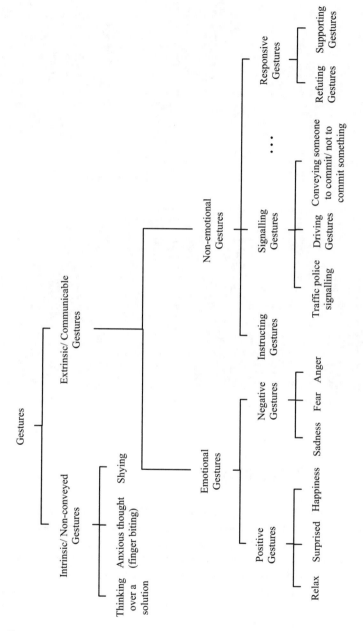

**Fig. 1.2** Gesture classification

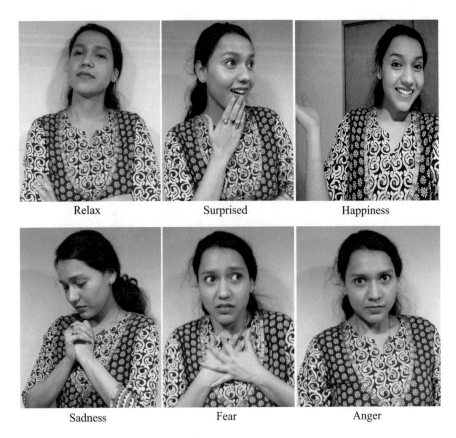

Fig. 1.3 Pictorial view of emotional gestures

## 1.2 Gesture Recognition

Gesture recognition refers to understanding and classifying gestures from human gestural data. Recognition of gestures usually is a pattern classification problem, where the individual and distinct gestures are classes and gestural instances containing gesture-data are instances. Both classical and machine learning approaches to gesture recognition are available. Among the traditional approaches, state-machines [3], Hidden Markov model (HMM) [4] and particle filter [5] need special mentioning. Machine learning approach, on the other hand involves designing a mapping function from gestural data to gesture-classes. Several machine learning techniques have also been employed [6, 7] to construct the functional mapping from gestural instances to gesture-classes. Since the classes are known a priori, gesture recognition is performed by supervised learning, where a supervisor fixes the training data set, describing gestural instances and the corresponding class

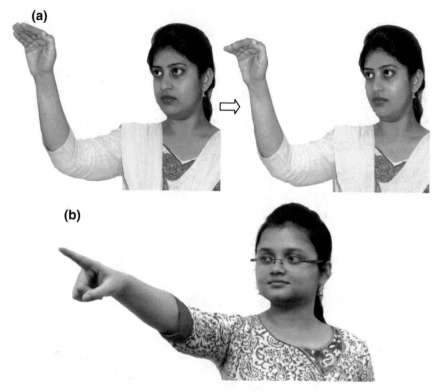

**Fig. 1.4 a** Pictorial view of instructing gesture "come here". **b** Pictorial view of instructing gesture "go there"

information for each gesture instance. After the mapping function is built from the gestural instances and class information, we can recognize an unknown gestural instance using the already available mapping function.

## 1.2.1   Gesture Recognition by Finite State Machines

A state machine [3] includes a set of states and a set of input symbols, where the state-transitions in the machine is controlled by the occurrence of a given input symbol at a given state. Let $\mathbf{S} = \{s_1, s_2, \ldots, s_n\}$ be a set of states and $\mathbf{A} = \{a_1, a_2, \ldots, a_m\}$ be a set of $m$ input symbols. Consider a deterministic state machine, where occurrence of an input symbol $a_k$ at a given state $s_i$ causes a transition in the machine to a fixed new state $s_j$. We use a nomenclature of delta functions $\delta(s_i, a_k, s_j) = 1$, if there exists a valid transition of state $s_i$ to $s_j$ for any input symbol $a_k$. It is indeed important to note that in a deterministic state machine, if $\delta(s_i, a_k, s_j) = 1$, there cannot exist any other symbol $a_l$, such that $\delta(s_i, a_l, s_j) = 1$. In a finite state

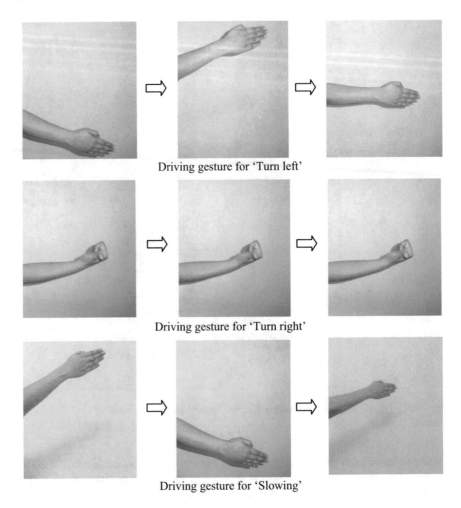

Driving gesture for 'Turn left'

Driving gesture for 'Turn right'

Driving gesture for 'Slowing'

**Fig. 1.5**  Pictorial view of driving gestures

machine, the cardinality of **S**, denoted by |**S**| is a finite number, i.e., the machine has limited number of states (Fig. 1.6).

A dynamic gesture can be represented by a sequence of primitive gestures. Usually, each primitive gesture is represented by a state. Thus a dynamic gesture is a collection of states, denoted by circles, where state-transitions, represented by directed edges from one state to the other, is invoked when a condition associated with a directed edge is satisfied. The conditions associated with an edge are connected by Boolean AND functions.

To make sense, let us consider the dynamic gesture of the arm movement of a person while attempting to sip a cup of tea. In Fig. 1.7, we present the states involved in the arm-movement gestural sequence of a subject, while taking a cup of

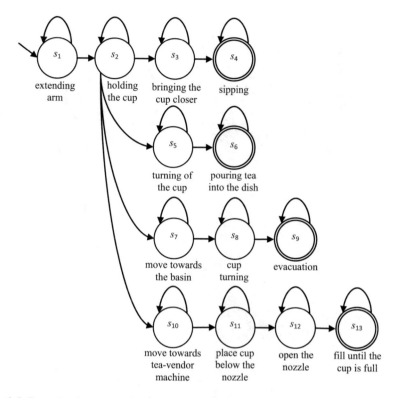

**Fig. 1.6** Example of state machine for four different dynamic gestures

tea. Four states are identified here as the primitive gestures. The primitive gestures include: (1) extending subject's arm towards the cup, (2) holding the cup by the forefinger with a support from the middle finger, (3) bring the cup close to the mouth, (4) sipping. The start state in the present context is extending one's arm towards the cup, and the goal state is sipping. State-transition rules here involve conditioning on the edges. Here, state-transition from state $i$ to state $i + 1$ is activated by the following two conditions.

*Condition 1* If the 2-dimensional centroid of an unknown predictable gesture is close to the same of a known gesture to a certain degree, then predictable gesture← selected known gesture.

*Condition 2* The transition delay between the current and the predictable gesture should be less than a user defined threshold.

For example, suppose the current state in Fig. 1.6 is state 2. After appearance of state 2, the computer looks for any one of four possible next states that may have transitions from state 2. In Fig. 1.6, the next state appears as state 3 satisfying conditions 1 and 2 both. The process of obtaining next state is continued until the sequence reaches the terminal state (here sipping).

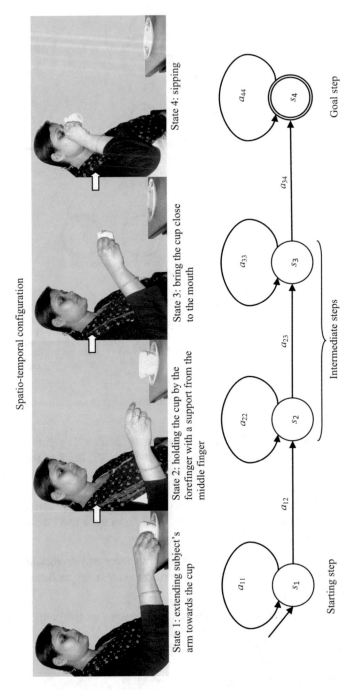

**Fig. 1.7** Fractional view of FSM for arm-movement gestural sequence of a subject while taking a cup of tea

It is noteworthy that a single state machine can recognize quite a few dynamic gestures. For example, the state machine, given in Fig. 1.6, can recognize four different gestures, representative of sipping a cup of tea, pouring tea into the dish, evacuation of the cup and filling the cup with tea. Figure 1.7 is a part of Fig. 1.6, as it describes a dynamic gesture of sipping a cup of tea.

## 1.2.2 Gesture Recognition by HMM

A HMM [4] is an extension of basic state-machines, where each state, denoted by a node, represents an event and the state-transitions are denoted by an arc function with a state-transition probability. The HMM also includes an output probability to designate the degree of the output variables that emanate from the machine. HMMs have successfully been applied for recognition of dynamic gestures, representing a sequence of gestural activities. Like a traditional Markov Chain [8], the HMM too determines the probability of the next state by only consulting the probability of the current state. Such single step state-transition probability is useful in many real world problems, concerning virtual gaming, motion control of robots, camera pan/tilt angle control and many others.

Like a Markov decision process, the state-transition in HMM is controlled by a power matrix. Let $\mathbf{W}$ be the state transition probability, whose columns denote the state-transition probabilities for different tasks/gestures, and rows denote time-locked gestures. We take $\mathbf{W}^m$ to obtain the state-transition probability from state $i$ to state $j$ in $m$ steps. Given a input vector denoting the probability of occurrence of each state at time $t$, we can obtain the probability of the next states at time $t + d$ by taking a product of the input vector by the $m$-th power of state transition probability matrix. This is given by

$$x_m = x_0 \mathbf{W}^m, \tag{1.1}$$

where $x_m$ and $x_0$ respectively denote the state-vector at time $t = m$ and $t = 0$. In traditional Markov chain, convergence of the $\mathbf{W}^m$ depends on the $\mathbf{W}$ matrix.

Two common observations follow from $\mathbf{W}$ matrix. If for a given $m$, $\mathbf{W}^m$ is a null matrix, then $x_m$ converges to a null vector, and the Markov system is said to have reached an absorbing state. If $\mathbf{W}^m = \mathbf{I}$, the identity matrix, then $x_{m+k} = x_0 \mathbf{W}^{m+k} = x_0 \mathbf{W}^k = x_k$, i.e., $x_{m+k} = x_k$ and the system demonstrates a limit cyclic (oscillatory) behaviour. There exist plenty of research results on the dynamic behaviour of the Markov matrix $\mathbf{W}$, the details of which are available in [8].

Recognising dynamic gesture using HMM requires segmenting the complete gesture into parts, where each part includes the basic primitives, so that the sequential union of which represent the entire gesture. We describe each primitive by a state and measure state-transitions by probability. The state-transition probabilities are defined in a manner, such that sum of these at a state $s$ due to a given input symbol $a$ is equal to 1. Determining the state-transition probabilities is a critical issue and thus needs subjective expertise/knowledge. Once this is prepared, we can use state-transition probabilities to determine the probability of recognizing a given sequence of input symbols, describing the gesture.

### 1.2.3  Gesture Tracking by Particle Filter

One fundamental issue in dynamic gesture recognition is to track the motion of the moving part of the human being. *Kalman filters* [9] and *Particle filters* [5] are commonly adopted techniques for tracking moving objects. In Kalman filter, we usually have a set of measurement variables, a set of estimator variables. A mathematical model of the estimator variables as a function of measurement variables is developed. Kalman filter employs a recursive algorithm to adapt estimator variables and an error covariance matrix in each step of recursion with an aim to reduce moving root mean square noise until the terminating condition is reached. The terminating condition usually adopted is to test the elements of the error covariance matrix. If all the elements of the error covariance matrix are very small, of the order of $10^{-3}$ or less, the recursion ends and the algorithm reports the converged estimator variables.

A common question that is apparent to readers is how Kalman filter differs from the traditional regression techniques. It is noteworthy that in traditional regression technique/curve fitting, we know the basic form of the estimators as functions of the measurements. There exist quite a vase literature [10, 11] to determine the unknown parameters in the estimator equations by least square regression techniques, where the motivation is to take the difference between the estimator response from the presumed model for known measurement variables and also to have knowledge about the estimators from the real system. We need to minimize the estimator error, which usually is performed by a least square estimation. However, the regression techniques take input data samples of measurements once for all. In a dynamic system, the parameters of the system change continuously over time. Thus traditional regression techniques are no longer useful for the present scenario. Here lies the importance of Kalman filter that takes input in succession and runs the recursive algorithm in a loop until the error converges, as evident from the error co-variance matrix.

Particle filters are of recent developments that outperform Kalman filter with respect to the following points. Kalman filter works fine for linear system model and convergence speed is also reasonably less. However, if the plant model is non-linear, the computations involved are extensive and thus not recommended for real-time application. Particle filter, however, does not have such restriction. A particle filter works well for linear/non-linear plant model. For non-linear plant model, we take samples of the system states and use an estimator using recursive Bayesian model. The recursive process used in a Kalman filter is optimal for linear plant model with Gaussian measurement noise. However, for non-linear system particle filter may yield better results at the cost of computational overhead.

The particle filter is similar to unscented Kalman filter (UKF) [12] by the following counts. In both UKF and particle filter, we have non-linear plant model and both the techniques estimate mean and co-variance of the system state. However, in particle filter the points are chosen randomly, while in UKF there is an attempt to select the points. This in turn requires selection of a large number of points in particle filter than in UKF, causing more computational overhead. Since number of particles is infinitely large, the estimator error converges to zero, unlike Kalman filter, where a small finite error remains even for large computational overhead.

The name 'particle filter' came after combining particle and filter. The term filter means to determine the distribution of a variable at a specific time. Particles mean weighted samples of the distribution. It uses recursive Bayesian filter by Monte Carlo sampling. Recursive filters are used as they allow batch processing; computation of all data can be done at one step. So the system becomes faster, allow on-line processing of data. In Monte Carlo sampling, particles are chosen randomly. With the increasing number of samples, convergence increases (Fig. 1.8).

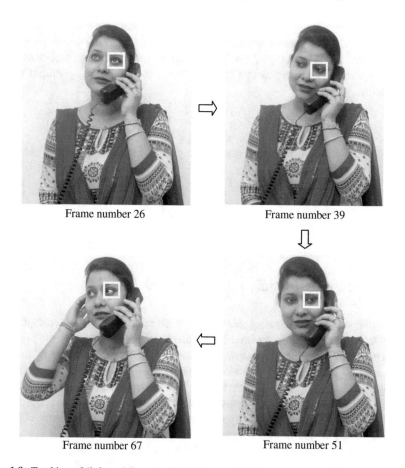

Frame number 26                    Frame number 39

Frame number 67                    Frame number 51

**Fig. 1.8**  Tracking of 'left eye' from a video using particle filter

Particle filter initializes an initial state value and then draws samples (let total number of samples are $N_p$) for the current state. Each sample $x_t^i$ ($i = 1, 2, ..., N_p$) denotes a hypothetical state of the tracked gesture with corresponding discrete sampling probability of $w_t^i$ at time $t$ such that

$$\sum_{i=1}^{N_p} w_t^i = 1. \tag{1.2}$$

The probability is found such that the particle is an accurate representation of the state (described by a vector $X_t$) based on the observation (described by a vector $Y_t$). Next state can be determined using posterior density $(X_t|Y_t)$ and posterior density $(Y_t|X_t)$. The state and measurement equations are

$$X_t = f(x_{t-1}, u_{t-1}, v_{t-1}). \tag{1.3}$$

$$Y_t = h(x_t, u_t, n_t). \tag{1.4}$$

where, f and h are the evolution and h measurement functions, which are usually not Gaussian. $u$ is known input, $v$ is the state noise and $n$ is the measurement noise.

Resampling of the weights can be done when required using posterior probability using

$$w_t^i = P(y_t|X_t = x_t^i). \tag{1.5}$$

In each iteration, of the algorithm, the weighted state of the gesture is measured as

$$E(S_t) = \sum_{i=1}^{N_p} w_t^i x_t^i. \tag{1.6}$$

## 1.2.4 Gesture Recognition by Supervised Learning

Traditional supervised learning algorithms generally employ a hyper-plane to segregate two sets of training instances representative of two classes of gestural data. Designing a suitable *hyperplane*, capable of correctly separating the data points of two classes, is an optimization problem. The aim of the optimization problem is to maximize the distance of the hyper-plane from the separating hyper-plane. Algorithms like *Linear Discriminant Analysis* (LDA) attempt to identify the parameters of the hyper-plane by maximizing the said objective function using classical calculus approach. In other words, the solution is obtained by setting the first derivatives of the objective function with respect to individual parameters of the hyperplane to zero. This effectively transforms the problem into a set of linear equations, which on solving return unique parameters for the hyper-plane.

Several variants to LDA are available in the literature [13, 14]. For instance, when the linear hyperplanes used in LDA cannot offer acceptable solutions (i.e., data points of two classes are not correctly separated), people go for *Quadratic Discriminant Analysis* (QDA), which employs a quadratic formulation of the hyperplane. Further, in order to ensure a significant margin between the two classes of data points, the formulation is extended further in *Support Vector Machines* (SVMs) [15]. SVMs also have two common variants, *Linear SVMs* (LSVMs) and *Kernelized SVMs* (KSVMs). In linear SVMs, the hyperplane is a linear function. In case data points of two classes are not linearly separable by adoption of LSVMs, we go for KSVMs, where the data points are projected to a high dimensional space by specially selected Kernel function, such as *Radial Basis Function* (RBF) Kernel and the like.

## 1.3 Uncertainty in Gesture Analysis

Gestural data of two classes can be separated by any one of the above formulations. However, to improve classifier performance, we need to eliminate noise that creeps into the gestural data for poor sensor characteristics and/or poor physical model employed to sense the gestural data. Here lies the importance of the book. We here adopted several sophisticated techniques to pre-process the acquired gestural data to eliminate contaminations from noise. In addition, we developed robust classifiers, which is insensitive to small Gaussian noise [16, 17].

To have a glimpse of the noise-tolerant classification, we here briefly demonstrate one specific characteristic of the logic of *fuzzy sets* [18, 19]. In the logic of fuzzy sets, we define a membership function to represent the degree of a linguistic variable $x$ to lie in a given set $A$, lying in a well-defined universe of discourse $X$. This degree of membership typically lies in [0, 1], where 0 denotes that $x$ does not belong to $A$, whereas 1 denotes that $x$ belongs to $A$ with a full degree of membership = 1, i.e., the occurrence of $x$ in $A$ is certain. Now, any positive membership other than 1, represents a partial membership of $x$ in $A$. This partial membership function (MF) maps the variable $x$ lying in the domain $[-\infty, +\infty]$ to [0, 1]. The nonlinear mapping cramps the small noise into smaller quantized values and is not visible in the membership, even if it is pronounced in $x$.

Thus logic of fuzzy sets is a good tool to handle small Gaussian noise in a linguistic variable $x$. When this characteristic is embedded in a classifier, it becomes robust and classifies noisy instances into true classes. Over the last two decades, there had been significant research contributions on extending this power of noise handling characteristics of fuzzy sets. A few well known contributions in this regard that need special mention are *type-2 fuzzy sets* [20–22], *intuitionistic fuzzy sets* [23], and many others. The merit of type-2 fuzzy sets lies in representing the uncertainty involved in the assignment of the memberships. For instance, let membership assigned for the linguistic variable height in fuzzy set MEDIUM at height = 5' be 0.9. This assignment suffers from *intra-personal* and *inter-personal uncertainty*. The intra-personal uncertainty refers to the degree of correctness in the assignment of the value 0.9 by the specific expert under reference, while the inter-personal uncertainty corresponds to difference in membership assignment at height = 5' by different experts. Fortunately, Type-2 fuzzy sets can handle both the above forms of uncertainty. In intuitionistic fuzzy sets (IFS), we instead of assigning only the degree of truth of a linguistic variable in a fuzzy set, also assign the degree of non-membership of the same variable in the same fuzzy set. IFS have good scope of uncertainty management as it carries the representational power of degree of truth and falsehood jointly, where the sum essentially need not be equal to one.

In this book, we primarily focus on to Type-2 fuzzy sets for pattern classification under uncertainty. Although type-$n$ fuzzy sets [24] are available to represent and model uncertainty in a better sense, because of high computational overhead to determine the type-$n$ memberships for $n > 2$, its use is restricted to type-2 fuzzy sets only. Two variants of type-2 MFs, called *General Type-2 Fuzzy Sets* (GT2FS) and *Interval Type-2 Fuzzy Sets* (IT2FS) are available. While both of these can handle inter-personal uncertainty identically; GT2FS can capture the intra-personal uncertainty in a better sense than its counterpart. The book contains a major section on GT2FS modelling for gesture recognition.

## 1.4    Gesture Learning by Neural Networks

Besides LDA, QDA and SVM, there exist quite a few interesting connectionist algorithms realized with Artificial Neural Network models for applications in pattern classification. A few of these that deserve mentioning includes *Back-Propagation Neural Networks* (BPNN) [25], *Radial basis Function Neural Networks* (RBF-NN) [26], *Probabilistic Neural Nets* (PNNs) [27], *Cascade Correlation Learning Architecture* (CCLA) [28]. All the above algorithms are meant for feed-forward neural architectures. Such architecture ensures signal propagation only in the forward direction.

The BPNN [25] algorithm is designed using the well-known *gradient descent learning* (GDL) algorithm. The GDL works on the following principle. It determines the next point at unit distance along the negative gradient of a given point and assigns the computed point to the current position on the surface in an iterative manner until the computed next point has zero slope. This Newtonian mechanism of GDL introduced above has been employed in a feed-forward neural network to adapt weights in BPNN algorithm. The BPNN algorithm adopts the above policy to adjust weights layer-wise, beginning with the last layer of the feed-forward topology. For a $n$-layered network, we need to have weight adjustment at $n - 1$ levels, which together is called a pass. After each pass, we compute the error vectors at the output layer for each pair of input-output instances, and repeat the training process with *Root-Mean Square* (RMS) error measure of all the input-output instances. Several tens of thousands passes are required before the network converges. On convergence, we record the network weights for subsequent recall actions for unknown stimulus

In an RBF neural network [26], we employ $m$ number of RBF neurons, where each neuron executes an RBF function to measure the similarity between an unknown input vector instance and the information stored at the RBF neurons. In case the input vector tallies with the stored information in a neuron, the neuron responds with an output = 1. Thus for $n$-class classification problem, we need $n$ RBF neurons, where each neuron is pre-assigned with an input vector describing the centroid of the pattern class to which it belongs. As a consequence, an unknown input instance can be correctly classified to the desired class. The RBF neural network comprises three layers, the input layer, the RBF layer comprising RBF neurons and the output layer, offering binary decisions about the unknown instance of each class. The last layer offers decisions like traditional *Perceptron neural networks*.

BPNN usually requires large training time. Besides, the neuronal weights in BPNN have a tendency to get trapped at local minima. PNN [27], however, overcomes the above limitations, and thus is ideally suited for supervised learning applications. A PNN realizes the well-known *Kernel Discriminant Analysis*, in which the different actions are arranged in layers of a feed-forward neural network (FFNN). The structure of a PNN is very similar to a BPNN. Only the sigmoid activation function is replaced by Bayes decision strategy to ensure convergence, irrespective of the dimension of the training set. The most prominent advantage of PNN over BPNN lies in the provision of augmenting training instances without disturbing the already adapted weights of the existing neurons.

As already indicated, BPNN suffers from large computational overhead, which increases too with the dimension of the network and that of the training data. In this book, the above limitation of BPNN has been overcome by employing CCLA [28]. In CCLA, the network initiates with a small structure having fewer input and output nodes. When the error term at an output node exceeds a pre-defined threshold, the network starts growing by introducing hidden layer to reduce error. All the hidden neurons have fixed weight in the input side and only the output side weights are trained. In other words, the CCLA trains the weights directly connected to the outputs, which results in incredible speed-up. Training the weights directly connected to the output neurons in CCLA, motivates employing single-layered learning algorithms, such as the Widrow-Hoff delta learning rule, the Perceptron learning and/or the resilient propagation (RPROP) algorithm.

## 1.5 Composing of Dynamic Gestures

In many applications, we occasionally need to match a gestural pattern portrayed in one or more frames of a video with standard gestural primitives. Searching a gestural primitive in a movie is not easy as it requires exponentially large computational overhead. In this book we adopt *Evolutionary Algorithms* (EAs) [29] to handle the situation.

Evolutionary algorithms (EAs) are population-based search algorithms, which aim at optimizing a given objective function. Several EAs have been developed over the last three decades, each with a primary motivation of capturing the optima by specialized genetic operators. For instance, in *Genetic Algorithm*, we employ three basic operators, called crossover, mutation and selection. While crossover and mutation are used for variation in the population (trial solutions), selection is used to select better offspring from a combined parent plus children population. In *Differential Evolution* (DE) algorithm [30], we employ crossover and mutation operations for variation in population and selection to identify the best-fit solution among the generated trial solutions generated in the current iteration.

A dance-video comprises a number of image frames. In composing a dance from $n$ videos, we need to check whether each consecutive pair of image frames has structural similarity and consistency. One simple approach to develop a new dance is to join $m$ videos in a manner, such that the last frame of the preceding video and the first frame of the consecutive video have structural similarity. The similarity of two videos is measured by a parameter, called *transition non-abruptness*. Thus when $m$ videos are ordered in a fixed sequence, we measure the transition non-abruptness of each two consecutively used videos and measure the sum of the possible transition non-abruptness. Since selection of m videos can be done in $n^m$ ways, we employ an evolutionary algorithm to minimize the fitness, representing the sum of transition non-abruptness in the resulting video. The book includes one chapter on dance video composition by evolutionary optimization.

## 1.6   Gesture Acquisition Techniques

Gesture acquisition typically is performed on images or by specially designed sensors involving camera/ultrasonic/*infra-radiometric* (IR) sensors. Ultrasonic sensors having wider spreading are used in very restricted applications. Infra-radiometric sensors operating at larger wavelength have less path-deviation [31] and thus is a good choice for range estimation of human gestural information. Alternatively, binocular vision, realized with two cameras in visible wavelength also is a good choice for range measurements in postural information.

Microsoft developed a commercially available instrument, called Kinect machine that captures the gestural information compositely from the camera information and infra-radiometric sensors. The infra-radiometric sensors include an infrared source and infrared detectors. Typically, the source is realized using a laser/ Light Emitting Diode (LED). Further, the detector is realized with a PIN diode. Such source-detector pairs employ *time-of-flight* principle. The time-of-flight principle determines the time required to traverse the infrared rays two times, once during forward journey and the other during return of the signal after reflection

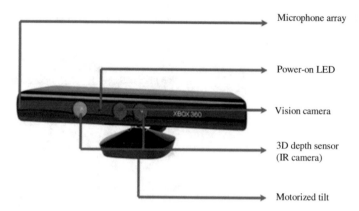

Microphone array

Power-on LED

Vision camera

3D depth sensor
(IR camera)

Motorized tilt

**Fig. 1.9**  Kinect sensor

from the object-surface. The difference in time-interval is used to measure the approximate distance of the object from the IR source. Microsoft Kinect.

Figure 1.9 provides a schematic diagram of the Microsoft's Kinect machine [32]. It is apparent from the figure that the machine includes both Infrared (IR) camera and one RGB camera. It also includes a microphone array to capture voice of the experimental subject. The IR camera and the RGB camera are used jointly to capture twenty body junction coordinates of the subjects including their depth as indicated in Fig. 1.10. To recognize a gesture, we pre-process the gestural data, then extract necessary features and finally classify the gesture from the noise-free/noise-reduced data.

Based on the sensory resolution, Kinect sensors have an output video frame rate of about 9–30 Hz. An 8-bit VGA resolution of 640 × 480 pixels with a Bayer color filter is used by the RGB videos. However, the hardware can also be used for resolutions up to 1280 × 1024 pixels and other formats such as UYVY. A VGA resolution of 640 × 480 pixels having 11-bit depth is possessed by the monochrome depth sensing video stream. This provides 2048 levels of sensitivity.

The origin of the reference frame is considered at the RGB camera. The direction parallel to the length of Kinect is the $x$-direction, the vertical direction is the $y$-direction and the $z$-direction is the depth measure explained in Fig. 1.11. Kinect records at a sampling rate of 30 frames per second (the sampling rate can be altered).

The Kinect sensor also has the ability to detect facial expressions and voice gestures. The pros of using the Kinect include its optimum cost, its ability to perform throughout the twenty four hours of the day. Other than these, the Kinect also reduces the problems that develop in the images of objects due to shadows and the presence of multiple sources of illumination. The range of distances within which the subject should perform in front of the Kinect is approximately 1.2–3.5 m

**Fig. 1.10** Twenty body joints

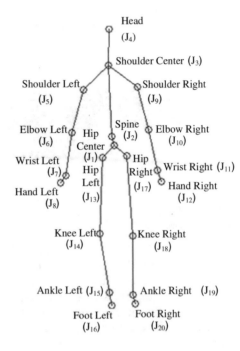

**Fig. 1.11** Three directions of reference frame

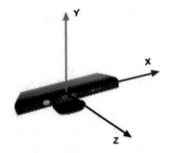

or 3.9–11 ft. With an increase in distance from the Kinect, the depth resolution decreases by about 1 cm at 2 m distance from the Kinect. The RGB data and the depth data together are required for visualization, which basically appears as a cloud around the joint, from which the features are being extracted.

Figure 1.12 describes the entire experimental setup used for dataset preparation by the authors. A white curtain is used in the background so that unwanted complexity of the background objects can be easily avoided. Two standing lights have also been kept to produce proper amount of light on the subject, so that the RGB images captured using camera become bright.

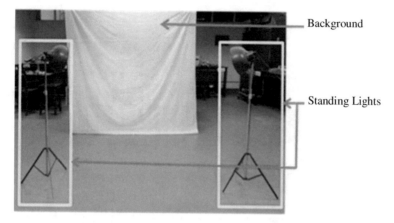

**Fig. 1.12** Setup for capturing training/testing dataset

## 1.7   Existing Approaches in Gesture Recognition

Gesture recognition can be done using two different modalities: two-dimensional (2-D) and three-dimensional (3-D) approaches [33]. The first technique is mostly done using image processing and further based on computer vision methods to obtain features from the 2-D shapes of the human body parts. Contrarily, 3-D human body recognition schemes deal with kinematics analysis, i.e. by calculating motions of concerned body parts by either intrusive or non-intrusive manners. Intrusive technique deals with placing markers on human body for tracing the movement of it, whereas nonintrusive methods rely on sensory data. In all the applications of gesture recognition, nonintrusive approaches are preferred over intrusive ones for better feasibility.

The major part of 2-D approaches mainly focuses on hand recognition for sign language and human-robot interaction applications. The first step (can be termed as pre-processing) deals with segmented the hand region from the background image. This is normally achieved using background subtraction and skin color based segmentation techniques. This process is followed with morphological operations with remove noise from the extracted image as depicted in Fig. 1.13. Feature extraction stage is based on calculating the shape, motion and/or position of the hand in the image.

To obtain the features, one of the methods is to superimpose a grid on the hand region in the image. From each grid, a possible number of features are computed that can explicitly define the state of the hand movement at time $t$. In [34], authors have proposed a similar technique of super-imposing a grid on the extracted hand gesture for sign language applications (Fig. 1.13). The extracted features from each grid can be of different types. Polana and Nelson [35] employ sum of the normal flow, Yamato et al. [36] calculate number of foreground pixels and Takahashi et al.

Proposed algorithm in [34]

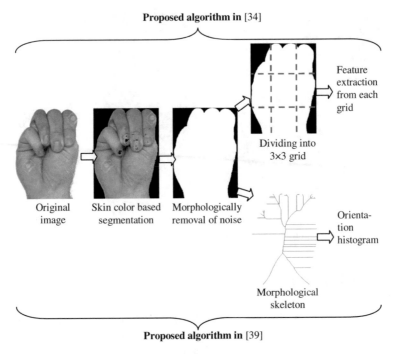

Fig. 1.13  Hand gesture recognition for American sign language alphabet 'M'

[37] implement an average edge vector for each grid. One of the advantages of using grid based representations is that it can handle the problem arising from partial occlusion.

For shape detection of hand, Freeman and Roth [38] propose *orientation histogram* based approach. Hariharan et al. [39] implement this technique for hand gesture recognition for an Indian classical dance form Bharatanatyam. Here two level of feature vector are computed, one is determining the *morphological skeleton* from the hand boundary and the other on is plotting the edge orientation histogram. Hunter et al. [40] executed Zernike moments to tackle rotation variant problem. In general, moments describe numeric quantities at some distance from a reference point or axis. Zernike moments projected the image function onto different orthogonal basis functions. Quek [41] state that when the hand is in motion, then movements of individual fingers can be neglected. In other words, hand gestures involving movements of finger with respect to each other is performed with small overall hand movement. Cootes et al. [42] propose a statistical shape model for detection and tracking of hands. Features are estimated based on principal component analysis (PCA) [43]. The paper lacks where occlusion occurs, i.e. when features are not present in the image. Some well-known hand gesture recognition techniques are tabulated in Table 1.1.

**Table 1.1** Existing literatures on hand gesture recognition

| Authors | Extraction technique | Recognition algorithms | Summary |
|---|---|---|---|
| Fels and Hinton [44] | Data glove | Multilayer neural network | Mapping of hand movements with speech using speech synthesizer using minor variations of the standard BPNN |
| Wilson and Bobick [45] | Data glove | Principal curve and dynamic time warping (DWT) | Extraction of motion trajectories to recognize various hand gestures in limited background |
| Starner et al. [46] | Colored glove | HMM | Recognition of continuous ASL by single camera using color-based tracker |
| Black and Jepson [47] | Colored marker | Modified condensation algorithm | Modelling of hand motions using temporal trajectories and matching using condensation algorithm by drawing random samples from the trajectories in office environment |
| Lien and Huang [48] | Vision based model | Inverse kinematics | Finding skeletal model for an input image by translating and rotating the hand model about $x$, $y$ and $z$ axes |
| Lee and Kim [49] | Skin color segmentation | HMM | Thresholding to calculate likelihood threshold of an input pattern and matching with known gesture patterns using confirmation mechanism |
| Yoon et al. [50] | Skin color segmentation | HMM | Tracking of hands in a gestural sequence by locating hand regions and connecting them for all frames. Features are based on weighted position, angle and velocity |
| Vogler and Metaxas [51] | 3-D computer vision or magnetic tracking system | HMM | Breaking of ASL into constituent phonemes and modelling them with parallel HMM |
| Yang et al. [52] | Multiscale segmentation | Time delay neural network | Computing affine transformations to acquire pixel matches and obtaining pixel-level motion trajectories using concatenation |

(continued)

**Table 1.1** (continued)

| Authors | Extraction technique | Recognition algorithms | Summary |
|---|---|---|---|
| Ng and Ranganath [53] | Skin color segmentation | Hidden Markov model and recurrent neural network (RNN) | Representing hand shapes by Fourier descriptors, calculating pose likelihood vector by RBF network and classifying pose using combination of HMM and RNN |
| Bretzner et al. [54] | Multiscale color features | Particle filtering | Detecting and tracking of hand states for controlling consumer electronics |
| Chen et al. [55] | Skin color segmentation | HMM | Extracting spatial features using Fourier descriptor and temporal by motion analysis |
| Chen et al. [56] | Background subtraction | Adaptive boosting (AdaBoost) and stochastic context free grammar | Recognition of hand posture using Haar-like features and Adaboost learning algorithm and hand gesture using stochastic context free grammar |
| Stergiopoulou and Papamarkos [57] | Skin color segmentation | Self-growing and self-organized neural gas (SGONG) network | Extraction of palm morphologic features based on the output grid of neurons and recognition using likelihood-based classification |
| Alon et al. [58] | Skin color segmentation | DTW | Recognition of hand gestures in complex backgrounds, which are vague in nature (location of hand and duration of the gesture), when foreground person performing the gesture |
| Suk et al. [59] | Skin color segmentation | Dynamic Bayesian network (DBN) | Recognition of continuous gestures for sign language recognition |

Liu et al. [60] design an algorithm uWave based on the principals of DTW. Here data acquisition is done using a single three-axis accelerometer. The authors have created a gesture vocabulary for this purpose. In another paper, Zhang et al. [61] propose an approach to recognize Chinese sign language using accelerometer and electromyography sensors. Fusion of decision tree and HMM is utilized to get the desired results. Template matching based algorithm to recognize hand gesture using MEMS accelerometer is developed by Xu et al. [62]. Another way of recognizing hand gesture is by representing them using depth maps by Kinect sensor [63]. This paper employs Finger-Earth mover's distance for identification of different hand shapes (while depicting different gestures) and claims to outperform the skeleton based approaches.

The identification of human body in images (2-D) is usually done by subtracting the background from the image. For these purpose, two major constraints are assumption of static background and a non-movable camera. Another criterion is self-occlusion, which makes it even more problematic to recognize haphazard movement of the humans. To model human body stick figure and ribbon technique find quite importance in the earlier works [64]. A ribbon based approach to recognize moving objects in 2-D space using continuity and proximity is proposed in [65]. Kahn et al. [66] propose a scheme to recognize static pointing posture from images, keeping in mind the texture of background, floor and lighting in the room for segmentation of the human body. Freeman et al. [67] developed an artificial retina chip for implement a low cost high speed for computer gaming. Based on image moments, an equivalent rectangle for the input image is created. Next, orientation histogram is used to calculate $x$ and $y$ derivatives from the input image for template matching. For better tracking from images, multiple cameras are preferred over a single camera [68]. Here the problem lies in matching the same object from different camera views. The authors have used Bayes' rule using Mahalanobis distance.

A shape representation using B-splines for tracking of pedestrians (assuming a static camera) is proposed by Baumberg and Hogg [69]. Here, a training shape is represented by a closed uniform B-spline with a fixed number of control points equally spaced around the boundary of the object. Spatio-temporal tracking is achieved by Kalman filtering in [70]. Another approach for pedestrian tracking is implemented using *wavelet co-efficients* (here Haar is taken) in [71]. The recognizing phase is consists with SVM to detect the frontal and rear views of the pedestrians. Wavelet analysis is essentially an extension of Fourier analysis which studies the coefficients of wavelets instead of those of sines and cosines. Haar wavelet is an orthogonal wavelet transform which uses a combination of averages and differences to study the details of a signal at many levels. Pedestrian detection involves shifting windows of various sizes over the image, normalizing Heisele et al. [72] use color of the pixels as the basis of tracking. Pixels are clubbed together in colors (R, G and B) and spatial dimensions $(x, y)$ using *k-means clustering* algorithm. The work does not guaranty that the tracking results will be locked into the same physical entity. The comparison of the outcomes of three algorithms is provided in Fig. 1.14.

The 3-D gesture recognition relies on some sort of sensor for detection of human body from the environment. It can be noted that limbs of animals does not move arbitrarily. This type of motion is coined as *biological motion* [73]. Based on this concept a human body can be modeled in a skeletal form. In a similar work, 3-D gait recognition from image sequence using stick-figure diagram is provided in [74]. Blob based tracking of moving persons is given in [75]. Campbell and Bobick [76] propose a 3-D scheme (using markers) for gesture recognition from nine ballet steps involving dynamic movement of the dancer. But here all the nine cases, the initial posture of the dancer is fixed. Euler angle to represent degrees of freedom in

**Fig. 1.14** Results of
pedestrian tracking using
three different algorithms

**Result obtained based on contour tracking algorithm** [69]

**Result obtained based on wavelet co-efficients** [71]

**Result obtained based on color cluster flow** [72]

XYZ plane and correlation is used to calculate fitness function of the dynamic
sequence. Sidenbladh et al. [77] develop a 3-D tracking scheme from 2-D gray scale
image sequence using Bayesian model having prior probability distribution over
pose and joint angles. Here, posterior probability distribution is propagated by
particle filtering over time using. But the application of the work is very limited to
tracking of arm and gait. Kang et al. [78] propose an algorithm for providing natural
interface to the gamers by monitoring continuous gestures. Here, skin color based
segmentation is used to extract three upper body parts (face and two palms). For
labeling of the frames, $k$-means clustering method is used and recognition stage
consists of DWT.

In a day-to-day life, most of the gestures are dynamic in nature. To recognize these types of gestures, HMM plays a significant role. Rigoli et al. [79] propose a method to recognize 24 gestures using HMM based on Viterbi algorithm. Here, feature extraction step measures a vector for each image frame based on the difference of the pixel intensity values with its previous image frame from the image sequence of a gesture.

In [80], a parametric HMM is designed where Baum-Welch algorithm is used in the training phase and in the testing phase a state is chosen where maximum probability is obtained. This algorithm is used to interpret parameterized gesture, i.e. gesture which has systematic spatial variation. A non-parametric model using left-right HMM is proposed in [81]. Here, the sequence of body poses is recognized by imposing temporal constraints among distinct poses. The approaches of gesture recognition rarely include hidden states and prediction of label for an entire gestural expression. To solve this problem, Wang et al. [82] implemented a hidden state conditional random field model both as one-versus-all and multi-class variant. Another tricky task is to recognize gesture from the whole body of the subject. In most of the research works, only hand and head gestures are recognized by segmenting based on skin color from 2-D images. But in [83], authors have proposed a method to spot whole body gesture in 3-D. Here angle of a particular body joint with respect to $x$, $y$ and $z$ co-ordinates are taken as inputs and sculpting of features are done using Gaussian mixture model. Recognition is done using HMM to interact with a robot based on the detected gesture. A comparative result is produced in [84] between SVM, HMM and conditional random field.

## 1.8 Scope of Gesture Recognition

Gesture recognition covers wide application-areas, ranging from development of medical instruments to entertainment purposes. Among the well-known medical applications, surgical robots need special mention. In a surgical robot, a senior physician, suffering from fatigue due to overwork, instructs a robot arm to serve as a surgeon. The instructions are usually provided by gestures. Assisted robotic surgery is also very common in many hospitals in the west, where the robots act as assistants to help the doctor with medicines, instruments and even cottons, as and when needed. These robots understand the doctor's need based on his gestural commands. Among other medical applications, patient monitoring system using Microsoft Kinect is very challenging. In patient-monitoring system, one or more Kinect machines are employed to monitor the health condition of the aged people at home. For instance, to monitor patients suffering from arthritic diseases, specific gestural signatures, resembling different arthritic symptoms, are identified by the physicians; a Kinect based gesture-recognition system is introduced next to recognize the specific gestural pattern of the patients for subsequent hospitalization in case of emergency.

Recent advances of sensors and human motion analysis have enabled the robots with improved perceptual abilities. Employment of robots have become such pervasive these days that the leading corporations are launching highly powerful robots along with efficient interfaces to re-establish their position in the market. Gradually, the involvement of robot manipulative end-effector in work environment accomplishing different tasks became so prominent that it opened an entire new dimension in gesture based robotic research. Also for noisy environments in several factories, gesture recognition plays an important role for signalling between human-to-human and human-to-robot.

With the widespread use of computers, sign language interpretation has emerged as one of the most promising research sub-disciplines in the vast domain of hand gesture recognition. The importance of sign language is apparent as the limitations of the traditionally used interfaces involving mouse and keyboard became more and more pronounced. A sign language typically refers to a way of interaction by visually transmitting the sign patterns embodied by a subject to convey an intended message. It is an efficient communication tool for speech and hearing impaired people; so the main motivation for developing such a system lies in the contribution of the system towards increasing social awareness and worth of it in providing social aid. The main objective of sign recognition is to build a robust and reliable framework of transcribing sign languages into text or speech which in turn facilitates efficient communication between deaf and dumb people. Sign language recognition is an ill-posed problem because sign languages are multichannel, conveying different meanings at once, moreover, they are highly inflected that is the appearance of the same sign is likely to take different appearances depending on the subject enacting it. By recognizing automatic gestures for sign language, a gesture to speech synthesizer can be developed for deaf and dumb people.

Computer gaming is fascinating for a lot of users because of its entertainment quotient. Being a fundamental part of the computer based amusement and refreshment; there has been a complete transformation in the building procedure of such games. By addition of interesting features, the scientists are investing their efforts in order to make the gaming environment more realistic and enjoyable. Further, along with the entertainment aspect, it is extremely important to address the issue of flexibility, the users must be comfortable while playing such games and it should not be physically or mentally exhausting as well. To implement these developments, gesture tracking seems to be the most suitable way for driving based games that enables a user to drive virtually by vicariously holding an imaginary steering wheel and moving limbs to proceed through the game using gestures only. Also for real time gaming (e.g. cricket, golf etc.); if gestures of the gamer can be recognized by the computer itself then decision making during the game can be easy. It can be useful for training enthusiastic players by eminent coaches by e-learning.

With the recent advent of internet, e-learning program has emerged as one of the most popular internet accessory, which requires a certain human being to acquire different skills online with the help of internet, which wouldn't have been possible for the user to learn in person otherwise. It is needless to say that machine intelligence plays an important role in these kinds of systems. A novel cost effective and fast e-learning procedure of detecting different dance postures can be implemented so that a novice can be able to learn from those postures and correct himself in case of any wrong gesture.

In real-world applications, most of the actions between multiple persons are rarely periodic and most of them are mostly performed by aperiodic motions of the body (e.g. 'punching', 'greeting' etc.). These interactions between more than one people need to be monitored for a number of useful applications, e.g. automatic recognition of vehement activities in smart surveillance systems. Involuntary alarm can be generated to the nearby police station in case of detection of suspicious activities of the persons while monitoring subjective action any environment (e.g. in a parking lot).

## 1.9  An Overview of the Book

The book includes 9 chapters, including one introduction, one concluding chapter and one 7 original contributions by the authors and their research team.

This chapter provides an introduction to gesture recognition. It begins with a thorough review of the principles of gesture recognition. Various tools and techniques relevant to image processing, pattern recognition and computational intelligence, which have necessary applications in gesture recognition, are also briefly explained here. The chapter then elaborately introduces the Kinect sensor, which has recently emerged as an important machinery to capture human-gestures. This provides a through outline of the existing works on gesture recognition. The chapter sketches possible applications of gesture recognition. The scope of the book is also appended at the end of the chapter.

Chapter 2 purposes an approach to recognize automatic ballet dance posture by recognizing the similarity between a known ballet primitive with an unknown posture. The initial stage of recognition is separating the dance posture from the background. This is achieved using skin color based segmentation. In the next stage morphological skeleton operation is applied to generate stick figure diagram of the posture. The formed skeleton structure is burdened with insignificant anomalies, which are removed using morphological spurring operation to obtain the corresponding minimized skeleton. All the twenty ballet primitives are gone through all these steps and then they are grouped based on Euler number. The feature extraction stage deals with application of Radon transform on the minimized skeletons to generate line integral plot. In the testing phase, the above stated procedure is carried on the unknown posture and its corresponding Euler number and line integral plot is determined. Now this unknown line integral plot is matched with the line integral plots of the known postures which have the same Euler number using correlation technique. Based on correlation co-efficients, correctness of the unknown posture is determined.

Chapter 3 provides another method to recognize ballet postures using fuzzy logic. The significance of this work is that when a novice dancer is going to pick up ballet moves using e-learning, then there will be significant amount of uncertainties present in her posture with respect to the danseuse's posture. To tackle this problem, earlier approach is not going to give fruitful results. Thus the chapter starts with providing with a brief overview to fuzzy sets. The subsequent part of the chapter the detailed procedure of application of chain code to generate straight line approximated skeleton from the minimized skeleton is given. The matching of unknown posture with the known ballet primitives is the highlight of this chapter.

The above two procedures take an image of a dancer as an input. A color-sensitive segmentation technique is employed to the input image to recognize the human body from the background. However, because of non-uniformity in image intensity and color, it is usually hard to segment an object from the background. Here lies the importance of Chap. 4, where we instead of images adopted a sophisticated technique to capture gestures of subjects using a Microsoft Kinect machine. This device is aptly suitable for realizing video game consoles where hands-free control gives better user satisfaction to the gamers. For car driving games, users can get virtual experience of driving the car by directing the car based on their gestures, rather than pressing keys on keyboard. To achieve this, a Mamdani type-1 fuzzy inference system is executed which is based on two inputs (distance between two palm positions and the angle generated by the line joining two palms with the sagittal plane) from the gamers' gestures. The output of the system is rotation of the car in the virtual car racing game. The speed of the car, i.e., application of acceleration and brake for the car, is controlled using the relative displacement of the right and the left feet.

In Chap. 5, a vital application of gesture recognition, home monitoring for elderly people is taken into account. It is noteworthy that gestures of different persons for the exact same pathological disorder is not exactly same, rather it suffers from high level of uncertainties. This wider variation in the gesture space motivated us to apply a type-2 fuzzy set based recognition scheme for detection of pathological disorder. This is the major contribution of Chap. 5. In the earlier part of this chapter, basic details about type-2 fuzzy set are provided. The novelty of this work lies in creating a secondary membership curve by a unique method. The amalgamation of primary and secondary membership curves gives us the type-2 fuzzy set. The disorder from an unknown gestural expression is identified by calculation of its proximity with known classes by type-2 fuzzy set.

For e-learning of dance, not only fuzzy sets, but neural networks as well play a great role. In Chap. 6, a novel gesture dependent e-learning of dance is proposed. Here, Kinect sensor is used to detect human skeleton in three-dimensional space. From these skeletons of the dancers, knowledge of geometry is applied to conceptualize five planes from each skeleton. These planes are constructed such that they signify major body parts while in motion. The classification stage deals with probabilistic neural network. The results of this proposed system is compared with existing techniques and it is remarkable that probabilistic neural network outperforms the standard classifiers for human skeleton matching in e-learning applications.

Based on the success of dance gesture recognition, we have introduced automatic composition of dance from already given dance sequences with the help of differential evolution algorithm in Chap. 7. For this purpose, a huge number of dance moves need to be acquired in the training phase. Next the fitness of each of these sequences is evaluated by calculating the inter-move transition abruptness with its adjacent dance moves. The transition abruptness is calculated as the difference of corresponding slopes formed by connected body joint co-ordinates. It can easily be inferred that the fittest dance sequence is where least abrupt inter-move transitions are present. The generated dance sequence is examined and it is noticed that the generated dance composition by the differential evolution algorithm do not have significant inter-move transition abruptness between two successive frames, indicating the efficacy of the approach.

A very interesting application of gesture recognition by fusing Kinect sensor based skeletal data with brain signals recorded by electroencephalography sensor for rehabilitative patients is provided in Chap. 8. Due to damage in parietal and/or motor cortex regions of a rehabilitative patient (subject), there can be failure while performing day-to-day basic task. Thus an artificial limb can assist the patient based on his brain signals and body gesture. To concretize our goal we have developed an experimental setup, where the target subject is asked to catch a ball while his/her brain (occipital, parietal and motor cortex) signals electroencephalography sensor and body gestures using Kinect sensor are simultaneously acquired. These two data are mapped using cascade correlation learning architecture to train a Jaco robotic arm. When a rehabilitative patient is unable to catch the ball, then the artificial limb is helpful for catching the ball. The proposed system can be implemented in other applications areas where we need to direct an artificial limb for assisting the subject in some locomotive tasks.

Finally Chap. 9 concludes this book while providing some future directions to gesture recognition.

# References

1. C. Darwin, *The Expression of the Emotions in Man and Animals*, vol 526 (University of Chicago Press, 1965)
2. S. Mitra, T. Acharya, Gesture recognition: a survey. IEEE Trans. Syst. Man Cybern. Part C Appl. Rev. **37**(3), 311–324 (2007)
3. T.S. Chow, Testing software design modeled by finite-state machines. IEEE Trans. Softw. Eng. **3**, 178–187 (1978)
4. L. Rabiner, B. Juang, An introduction to hidden Markov models. IEEE ASSP Mag. **3**(1), 4–16 (1986)
5. M.S. Arulampalam, S. Maskell, N. Gordon, T. Clapp, A tutorial on particle filters for online nonlinear/non-Gaussian Bayesian tracking. IEEE Trans. Signal Process. **50**(2), 174–188 (2002)

6. P. Hong, M. Turk, T. S. Huang, Gesture modeling and recognition using finite state machines, in *Proceedings of the 4th IEEE International Conference on Automatic Face and Gesture Recognition* (2000), pp. 410–415
7. M. Brand, N. Oliver, A. Pentland, Coupled hidden Markov models for complex action recognition, in *Proceedings of the 1997 IEEE Computer Society Conference on Computer Vision And Pattern Recognition* (1997), pp. 994–999
8. J.G. Kemeny, J.L. Snell, *Finite Markov Chains*, vol. 356 (van Nostrand, Princeton, 1960)
9. M.S. Grewal, *Kalman Filtering* (Springer, 2011)
10. N.R. Draper, H. Smith, E. Pownell, *Applied Regression Analysis*, vol. 3 (Wiley, New York, 1966)
11. R.C. Littell, *Sas* (Wiley Online Library, 2006)
12. E.A. Wan, R. Van Der Merwe, The unscented Kalman filter for nonlinear estimation, in *Adaptive Systems for Signal Processing, Communications, and Control Symposium 2000. AS-SPCC. The IEEE 2000* (2000), pp. 153–158
13. S. Theodoridis, A. Pikrakis, K. Koutroumbas, D. Cavouras, *Introduction to Pattern Recognition: A Matlab Approach.* (Academic Press, 2010)
14. R.O. Duda, P.E. Hart, *Pattern Classification and Scene Analysis*, vol. 3 (Wiley, New York, 1973)
15. J.A.K. Suykens, J. Vandewalle, Least squares support vector machine classifiers. Neural Process. Lett. **9**(3), 293–300 (1999)
16. D. Bhattacharya, A. Konar, P. Das, Secondary factor induced stock index time-series prediction using Self-Adaptive Interval Type-2 Fuzzy Sets. Neurocomputing **171**, 551–568 (2016)
17. P. Rakshit, A. Konar, S. Das, L.C. Jain, A.K. Nagar, Uncertainty management in differential evolution induced multiobjective optimization in presence of measurement noise. IEEE Trans. Syst. Man Cybern. Syst. **44**(7), 922–937 (2014)
18. L.A. Zadeh, Fuzzy sets. Inf. Control **8**(3), 338–353 (1965)
19. H.-J. Zimmermann, Fuzzy control, in *Fuzzy Set Theory—and Its Applications* (Springer, 1996), pp. 203–240
20. J.M. Mendel, D. Wu, Perceptual reasoning for perceptual computing. IEEE Trans. Fuzzy Syst. **16**(6), 1550–1564 (2008)
21. Q. Liang, J.M. Mendel, Interval type-2 fuzzy logic systems: theory and design. IEEE Trans. Fuzzy Syst. **8**(5), 535–550 (2000)
22. C. Wagner, H. Hagras, Toward general type-2 fuzzy logic systems based on zSlices. IEEE Trans. Fuzzy Syst. **18**(4), 637–660 (2010)
23. K.T. Atanassov, Intuitionistic fuzzy sets. Fuzzy Sets Syst. **20**(1), 87–96 (1986)
24. L.A.Z. Adeh, Calculus of fuzzy restrictions. Fuzzy Sets, Fuzzy Logic, Fuzzy Syst. Sel. Pap. by Lotfi A Zadeh **6**, 210 (1996)
25. A.T.C. Goh, Back-propagation neural networks for modeling complex systems. Artif. Intell. Eng. **9**(3), 143–151 (1995)
26. L. Yingwei, N. Sundararajan, P. Saratchandran, A sequential learning scheme for function approximation using minimal radial basis function neural networks. Neural Comput. **9**(2), 461–478 (1997)
27. D.F. Specht, Probabilistic neural networks. Neural Networks **3**(1), 109–118 (1990)
28. S.E. Fahlman, C. Lebiere, The cascade-correlation learning architecture, in *Advances in neural information processing systems* (1990), pp. 524–532
29. E. Zitzler, L. Thiele, Multiobjective evolutionary algorithms: a comparative case study and the strength Pareto approach. IEEE Trans. Evol. Comput. **3**(4), 257–271 (1999)
30. R. Storn, K. Price, Differential evolution—a simple and efficient heuristic for global optimization over continuous spaces. J. Glob. Optim. **11**(4), 341–359 (1997)
31. P.W. Barone, S. Baik, D.A. Heller, M.S. Strano, Near-infrared optical sensors based on single-walled carbon nanotubes. Nat. Mater. **4**(1), 86–92 (2005)
32. Z. Zhang, Microsoft kinect sensor and its effect. IEEE Multimedia **19**(2), 4–10 (2012)

33. D.M. Gavrila, The visual analysis of human movement: a survey. Comput. Vis. Image Underst. **73**(1), 82–98 (1999)
34. S. Saha, S. Bhattacharya, A. Konar, Comparison between type-1 fuzzy membership functions for sign language applications, in *2016 International Conference on Microelectronics, Computing and Communications (MicroCom)* (2016), pp. 1–6
35. R. Polana, R. Nelson, Low level recognition of human motion (or how to get your man without finding his body parts), in *Proceedings of the 1994 IEEE Workshop on Motion of Non-Rigid and Articulated Objects* (1994), pp. 77–82
36. J. Yamato, J. Ohya, K. Ishii, Recognizing human action in time-sequential images using hidden markov model, in *Proceedings of the 1992 IEEE Computer Society Conference on Computer Vision and Pattern Recognition, CVPR'92* (1992), pp. 379–385
37. K. Takahashi, S. Seki, E. Kojima, and R. Oka, "Recognition of dexterous manipulations from time-varying images," in *Proceedings of the 1994 IEEE Workshop on Motion of Non-Rigid and Articulated Objects* (1994), pp. 23–28
38. W.T. Freeman, M. Roth, Orientation histograms for hand gesture recognition, in *International Workshop on Automatic Face and Gesture Recognition,* vol 12 (1995), pp. 296–301
39. D. Hariharan, T. Acharya, S. Mitra, Recognizing hand gestures of a dancer, in *Pattern Recognition and Machine Intelligence* (Springer, 2011), pp. 186–192
40. E. Hunter, J. Schlenzig, R. Jain, Posture estimation in reduced-model gesture input systems, in *International Workshop on Automatic Face-and Gesture-Recognition* (1995), pp. 290–295
41. F.K.H. Quek, Eyes in the interface. Image Vis. Comput. **13**(6), 511–525 (1995)
42. T.F. Cootes, C.J. Taylor, D.H. Cooper, J. Graham, Active shape models-their training and application. Comput. Vis. Image Underst. **61**(1), 38–59 (1995)
43. I. Jolliffe, *Principal Component Analysis* (Wiley Online Library, 2002)
44. S.S. Fels, G.E. Hinton, Glove-talk: a neural network interface between a data-glove and a speech synthesizer. IEEE Trans. Neural Networks **4**(1), 2–8 (1993)
45. A.F. Bobick, A.D. Wilson, A state-based approach to the representation and recognition of gesture. IEEE Trans. Pattern Anal. Mach. Intell. **19**(12), 1325–1337 (1997)
46. T. Starner, J. Weaver, A. Pentland, Real-time american sign language recognition using desk and wearable computer based video. IEEE Trans. Pattern Anal. Mach. Intell. **20**(12), 1371–1375 (1998)
47. M.J. Black, A.D. Jepson, A probabilistic framework for matching temporal trajectories: condensation-based recognition of gestures and expressions, in *European Conference on Computer Vision* (1998), pp. 909–924
48. C.-C. Lien, C.-L. Huang, Model-based articulated hand motion tracking for gesture recognition. Image Vis. Comput. **16**(2), 121–134 (1998)
49. H.-K. Lee, J.-H. Kim, An HMM-based threshold model approach for gesture recognition. IEEE Trans. Pattern Anal. Mach. Intell. **21**(10), 961–973 (1999)
50. H.-S. Yoon, J. Soh, Y.J. Bae, H.S. Yang, Hand gesture recognition using combined features of location, angle and velocity. Pattern Recognit. **34**(7), 1491–1501 (2001)
51. C. Vogler, D. Metaxas, A framework for recognizing the simultaneous aspects of american sign language. Comput. Vis. Image Underst. **81**(3), 358–384 (2001)
52. M.-H. Yang, N. Ahuja, M. Tabb, Extraction of 2d motion trajectories and its application to hand gesture recognition. IEEE Trans. Pattern Anal. Mach. Intell. **24**(8), 1061–1074 (2002)
53. C.W. Ng, S. Ranganath, Real-time gesture recognition system and application. Image Vis. Comput. **20**(13), 993–1007 (2002)
54. L. Bretzner, I. Laptev, T. Lindeberg, Hand gesture recognition using multi-scale colour features, hierarchical models and particle filtering, in *Proceedings of the 5th IEEE International Conference on Automatic Face and Gesture Recognition* (2002), pp. 423–428
55. F.-S. Chen, C.-M. Fu, C.-L. Huang, Hand gesture recognition using a real-time tracking method and hidden Markov models. Image Vis. Comput. **21**(8), 745–758 (2003)
56. Q. Chen, N.D. Georganas, E.M. Petriu, Hand gesture recognition using haar-like features and a stochastic context-free grammar. IEEE Trans. Instrum. Meas. **57**(8), 1562–1571 (2008)

57. E. Stergiopoulou, N. Papamarkos, Hand gesture recognition using a neural network shape fitting technique. Eng. Appl. Artif. Intell. **22**(8), 1141–1158 (2009)
58. J. Alon, V. Athitsos, Q. Yuan, S. Sclaroff, A unified framework for gesture recognition and spatiotemporal gesture segmentation. IEEE Trans. Pattern Anal. Mach. Intell. **31**(9), 1685–1699 (2009)
59. H.-I. Suk, B.-K. Sin, S.-W. Lee, Hand gesture recognition based on dynamic Bayesian network framework. Pattern Recognit. **43**(9), 3059–3072 (2010)
60. J. Liu, L. Zhong, J. Wickramasuriya, V. Vasudevan, uWave: accelerometer-based personalized gesture recognition and its applications. Pervasive Mob. Comput. **5**(6), 657–675 (2009)
61. X. Zhang, X. Chen, Y. Li, V. Lantz, K. Wang, J. Yang, A framework for hand gesture recognition based on accelerometer and EMG sensors. IEEE Trans. Syst. Man Cybern. Part A Syst. Hum. **41**(6), 1064–1076 (2011)
62. R. Xu, S. Zhou, W.J. Li, MEMS accelerometer based nonspecific-user hand gesture recognition. IEEE Sens. J. **12**(5), 1166–1173 (2012)
63. Z. Ren, J. Yuan, J. Meng, Z. Zhang, Robust part-based hand gesture recognition using kinect sensor. IEEE Trans. Multimedia **15**(5), 1110–1120 (2013)
64. T.B. Moeslund, E. Granum, A survey of computer vision-based human motion capture. Comput. Vis. Image Underst. **81**(3), 231–268 (2001)
65. S. Kurakake, R. Nevatia, Description and tracking of moving articulated objects, in *Proceedings of the 11th IAPR International Conference on Pattern Recognition, 1992. Vol. I. Conference A: Computer Vision and Applications* (1992), pp. 491–495
66. R.E. Kahn, M.J. Swain, P.N. Prokopowicz, R.J. Firby, Gesture recognition using the perseus architecture, in *Proceedings of the 1996 IEEE Computer Society Conference on Computer Vision and Pattern Recognition, CVPR'96* (1996), pp. 734–741
67. W.T. Freeman, K. Tanaka, J. Ohta, K. Kyuma, Computer vision for computer games, in *Proceedings of the Second International Conference on Automatic Face and Gesture Recognition* (1996), pp. 100–105
68. Q. Cai, J.K. Aggarwal, Tracking human motion using multiple cameras, in *Proceedings of the 13th International Conference on Pattern Recognition*, vol 3 (1996), pp. 68–72
69. A.M. Baumberg, D.C. Hogg, An efficient method for contour tracking using active shape models, in *Proceedings of the 1994 IEEE Workshop on Motion of Non-Rigid and Articulated Objects* (1994), pp. 194–199
70. A. Blake, R. Curwen, A. Zisserman, A framework for spatiotemporal control in the tracking of visual contours. Int. J. Comput. Vis. **11**(2), 127–145 (1993)
71. M. Oren, C. Papageorgiou, P. Sinha, E. Osuna, T. Poggio, Pedestrian detection using wavelet templates, in *Proceedings of the 1997 IEEE Computer Society Conference on Computer Vision and Pattern Recognition* (1997), pp. 193–199
72. B. Heisele, U. Kressel, W. Ritter, Tracking non-rigid, moving objects based on color cluster flow, in *Proceedings of the 1997 IEEE Computer Society Conference on Computer Vision and Pattern Recognition* (1997), pp. 257–260
73. D.D. Hoffman, B.E. Flinchbaugh, The interpretation of biological motion. Biol. Cybern. **42**(3), 195–204 (1982)
74. Z. Chen, H.-J. Lee, Knowledge-guided visual perception of 3-D human gait from a single image sequence. IEEE Trans. Syst. Man. Cybern. **22**(2), 336–342 (1992)
75. A. Azarbayejani and A. Pentland, "Real-time self-calibrating stereo person tracking using 3-D shape estimation from blob features," in *Proceedings of the 13th International Conference on Pattern Recognition*, vol 3 (1996), pp. 627–632
76. L.W. Campbell, A.F. Bobick, Recognition of human body motion using phase space constraints, in *Proceedings of the 5th International Conference on Computer Vision* (1995), pp. 624–630
77. H. Sidenbladh, M.J. Black, D.J. Fleet, Stochastic tracking of 3D human figures using 2D image motion, in *European Conference on Computer Vision* (2000), pp. 702–718
78. H. Kang, C.W. Lee, K. Jung, Recognition-based gesture spotting in video games. Pattern Recognit. Lett. **25**(15), 1701–1714 (2004)

79. G. Rigoll, A. Kosmala, S. Eickeler, High performance real-time gesture recognition using hidden markov models, in *International Gesture Workshop* (1997), pp. 69–80
80. A.D. Wilson, A.F. Bobick, Parametric hidden markov models for gesture recognition. IEEE Trans. Pattern Anal. Mach. Intell. **21**(9), 884–900 (1999)
81. A. Elgammal, V. Shet, Y. Yacoob, L.S. Davis, Learning dynamics for exemplar-based gesture recognition, in *Computer Vision and Pattern Recognition,* vol 1 (2003), p. I
82. S.B. Wang, A. Quattoni, L.-P. Morency, D. Demirdjian, T. Darrell, Hidden conditional random fields for gesture recognition, in *2006 IEEE Computer Society Conference on Computer Vision and Pattern Recognition,* vol 2 (2006), pp. 1521–1527
83. H.-D. Yang, A.-Y. Park, S.-W. Lee, Gesture spotting and recognition for human–robot interaction. IEEE Trans. Robot. **23**(2), 256–270 (2007)
84. L.-P. Morency, A. Quattoni, T. Darrell, "Latent-dynamic discriminative models for continuous gesture recognition," in *IEEE Conference on Computer Vision and Pattern Recognition, CVPR'07* (2007), pp. 1–8

# Chapter 2
# Radon Transform Based Automatic Posture Recognition in Ballet Dance

The proposed system aims at automatic identification of an unknown dance posture referring to the twenty primitive postures of ballet, simultaneously measuring the proximity of an unknown dance posture to a known primitive. The proposed system aims at automatic identification of an unknown dance posture referring to the twenty primitive postures of ballet, simultaneously measuring the proximity of an unknown dance posture to a known primitive. A simple and novel six stage algorithm achieves the desired objective. Skin color segmentation is performed on the dance postures, the outputs of which are dilated and processed to generate skeletons of the original postures. The stick figure diagrams laden with minor irregularities are transubstantiated to generate their affirming minimized skeletons. Each of the twenty postures based on their corresponding Euler number are categorized into five groups. Simultaneously the line integral plots of the dance primitives are determined by performing Radon transform on the minimized skeletons. The line integral plots of the fundamental postures along with their Euler numbers populate the initial database. The group of an unknown posture is determined based on its Euler number, while successively the unknown posture's line integral plot is compared with the line integral plots of the postures belonging to that group. An empirically determined threshold finally decides on the correctness of the performed posture. While recognizing unknown postures, the proposed system registers an overall accuracy of 91.35%.

## 2.1 Introduction

Ballet is an artistic dance form, performed by unequivocal and highly formalized set steps and kinesics. The proposed body of work deals with automatic posture recognition of the various dance sequences of ballet. Recognition, modeling and analysis of the various postures comprising the dance form help an enthusiast to learn the art by intelligibly interacting with the proposed system. The motivation

© Springer International Publishing AG 2018
A. Konar and S. Saha, *Gesture Recognition*, Studies in Computational Intelligence 724, DOI 10.1007/978-3-319-62212-5_2

behind this novel work is to propose a system for automatic learning of ballet dance in a simple and flexible manner.

Substantial work has been done in the field of posture recognition; however, very few of these deal with posture recognition in specific dance procedures. The work discussed in [1] deals with the sequencing of dance postures using stick figure diagrams. However the dependence of the algorithm on motion sensor devices escalates the cost of the interactive setup. In addition to that, the algorithm deals exclusively with the leg movements of the performer and in the process neglects the specifications of the other body parts. Without considering the entire posture of a performer, capturing the essence of a dance form as elaborate as ballet is not possible.

In [2], Guo and Qian propose an interesting scheme for gesture recognition in 3D dance. The algorithm presented in this paper is different from the one proposed by Guo and Qian [2], by the following. Unlike Guo and Qian, the algorithm proposed in this chapter is concerned with 2D posture recognition of ballet dance postures and uses a simple camera instead of the binocular vision system (i.e., dual camera setup) that has been used in [2] for orthogonal projections. The incorporation of 2D images enhances the classification accuracy from 81.9 to 91.3%. The proposed algorithm successfully tackles the anomalies that might appear due to the difference in the body-structure of ballet dancers that Guo and Qian fail to address.

The authors in [3] propose a dance posture recognition technique with the help of which issues related to the texture of the dress, distance of the performer with respect to the camera and body structure of the dancer is effectively addressed. With the help of the proposed algorithm, the dancer is effectively extracted from the background in which (s)he is performing. Apart from the above mentioned procedures dealing exclusively with dance postures, there exists a comprehensive body of work dealing entirely with human motion analysis and posture recognition.

Analysis of human gesture from different perspectives is an important issue. This can be done using image sequence analysis [4] by representing the human body using stick figure diagrams [5]. In [6], 2D posture recognition is done using silhouette extraction. Here, human body is segmented from complex background using a statistical method, and matching of postures is performed using genetic algorithm. In [7], Feng and Lin segment the object from the background using background difference method. Next the extract features concerning object shape and area and use a Support Vector Machine classifier to classify postures in elderly health care system. In [8], activity monitoring of elders is proposed with the help of silhouette extraction. In both [7, 8], the work involves sequences of images, i.e., postures are recognized from video.

Two similar approaches of gesture recognition in ballet dance are reported in [9, 10]. Here, the researchers aim at approximating the curved lines of the skeletons by straight line approximation using chain codes. However, their algorithm requires additional registration formalities to place the RGB image of the dancer at the centre of the image area. This factor unfortunately is of important concern in learning ballet dance. In the present work, we modify the line integral plots of the skeleton of the dancer by rejecting leading and trailing zeros and thus the problem of centralizing does not appear. Also the accuracy of the proposed work is better than those of

[9, 10]. The proposed work deals with static posture recognition and thus cannot be compared with dynamic gesture recognition techniques introduced in [11, 12].

In the present work, we are dealing with 20 basic postures of ballet dance. RGB images pertaining to the primitive postures of ballet constitute the initial database. Unclassified ballet postures act as inputs to the algorithm. Skin color segmentation is performed on the input images, to get rid of the unnecessary background information. The extracted postures are modeled using skeletonization. Information derived from the stick figure diagrams include the Euler numbers of the skeletons and the line integral plots using Radon transform. Finally a sum of correlation coefficients measure is employed to classify the unknown dance posture. The information extracted is processed according to the proposed technique, which not only detects a posture but additionally determines its correctness. The preferred system makes learning ballet an interactive exercise whereby the learner interacts with the computer in an efficient as well as economical manner.

## 2.2 Methodology

This section explains the preprocessing steps and the principles used for posture recognition in great detail. The flowchart corresponding to the proposed algorithm is given in Fig. 2.1.

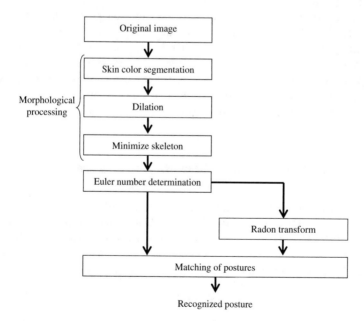

**Fig. 2.1** Block diagram of the proposed algorithm for automatic posture recognition in ballet dance

## 2.2.1  Morphological Processing

Color images of the ballet postures in RGB format are treated as inputs to the proposed algorithm. A particular dress pattern is recommended while performing ballet and owing to its specifications, a major portion of the dancer's body remains exposed. Hence, skin color segmentation is used to extract the performer from the background. The skin and non-skin pixels are distinguished from each other using uni-modal Gaussian model [13]. Let $m$ is the mean vector and $C$ is the co-variance matrix of skin color. Again, $x$ is a skin pixel if the following equation holds

$$(x - \vec{m})^T \vec{C}^{-1}(x - \vec{m}) - (x - \vec{m}_n)^T \vec{C}_n^{-1}(x - \vec{m}_n) \leq \tau \qquad (2.1)$$

where $m_n$ and $C_n$ are the mean and co-variance of skin and non-skin colors respectively. For our purpose, we have taken threshold value ($\tau$) as 0.01 [14].

On performing skin color segmentation, certain sections of the original RGB image are wrongly classified as skin segments. In order to get rid of this problem, connected components (each connected to its eight neighbors) are considered in the binary image and components having less than 100 pixels are automatically removed from it.

The skin color segmented images are dilated with the help of two line elements, each of length 3 units and corresponding to angles 0° and 90° respectively. The dilation procedure corrects the minor irregularities present in the segmented image to a considerable extent.

In the next step, the dilated images are skeletonized giving rise to stick figure representations of the concerned dance postures. The skeletons so formed contain certain irregular constructs, which add no valuable information to the image. As a result of which shorter length lines are eliminated from the medial axis images with the help of a morphological spur operation. The minimized skeletons finally contain the significant lines that maximally determine a dance posture. Figure 2.2 contains a step by step depiction of the entire process.

## 2.2.2  Euler Number Based Grouping of Postures

A total of 20 different postures are grouped into five different categories based on the Euler number [15] of the minimized skeletons of the corresponding postures. The generated skeletons contain variable number of open loop lines and closed loop holes. The Euler number ($E$) is determined after calculating the number of holes ($H$) and connected components ($CC$) in a particular skeleton.

$$E = CC - H \qquad (2.2)$$

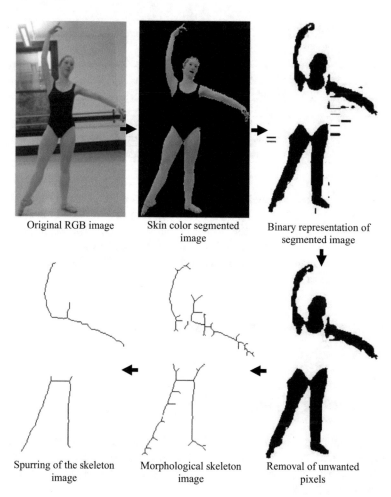

**Fig. 2.2** Results of different stages during morphological processing for 'Ecarte Devant' posture

The Euler numbers range from −1 to +3 and based on the number of connected components and holes present in the postures, they are categorized into individual groups (*G*). Table 2.1 presents the five different categories and the postures belonging to them, whereas Fig. 2.3 provides examples of postures belonging to each category.

## 2.2.3 Radon Transform

Skeletons belonging to the same group are differentiated with the help of Radon transform [16, 17]. Each pixel constituting the minimized skeletons is projected

**Table 2.1** Grouping of primitive postures

| Euler Number ($E$) | Groups ($G$) | Example |
|---|---|---|
| −1 | 1 | Arms First, Posture Front |
| 0 | 2 | Arms Third, Arms Fourth, Releve |
| 1 | 3 | Arms Fifth, Posture Side |
| 2 | 4 | Attitude Front, Croise Derriere, Croise Devant, Ecarte Derriere, Ecarte Devant, Efface Derriere, Efface Devant, En Face, Epaule |
| 3 | 5 | Arabesque, Arms Second, Attitude Back, Attitude Side |

along the $x$ and $y$-axes. Figure 2.4 represents the Radon transform, wherein the red color represents projections along the $x$-axis (i.e., along $0°$ angle) and the blue color represents projections along the $y$-axis (i.e., along $90°$ angle) respectively. Considering the image in the spatial domain $(x, y)$ mapped along the projection domain $(p, \theta)$, the required Radon Transform of a function $f(x, y)$ denoted by $R(p, \theta)$ can be expressed as

$$R(p, \theta) = \int_{-\infty}^{+\infty} f(x, y)dq = \int_{-\infty}^{+\infty} f(p\cos\theta - q\sin\theta, p\sin\theta + q\cos\theta)dq'. \quad (2.3)$$

### 2.2.4  Matching of Postures

The final stage of the proposed algorithm deals with recognition of an unknown ballet posture with the help of matching of their corresponding Radon transforms. An initial database is prepared by storing the radon transforms of the twenty basic dance postures of ballet.

The line integral plot of an unknown posture is determined and is matched with the line integral plots of all the postures belonging to the group to which the unknown posture belongs. In order to nullify the effect of height, weight and body structure of the dancer, the line integral plot of the unknown and known postures are scaled accordingly. For the purpose of scaling, the leading and trailing zeroes, which often form a part of the line integral plots in the two axes, are neglected and the pertinent portion of the plot is considered. This approach helps us to compare non-centralized images adding a lot of flexibility to the entire procedure. The number of points from relevant portions are either scaled up or down to ensure that the number of such points is a multiple of 50. In order to scale down, the number of points that needs to be deleted is selected at equal intervals. In order to scale up, the number of points that needs to be appended is added at regular intervals. The new

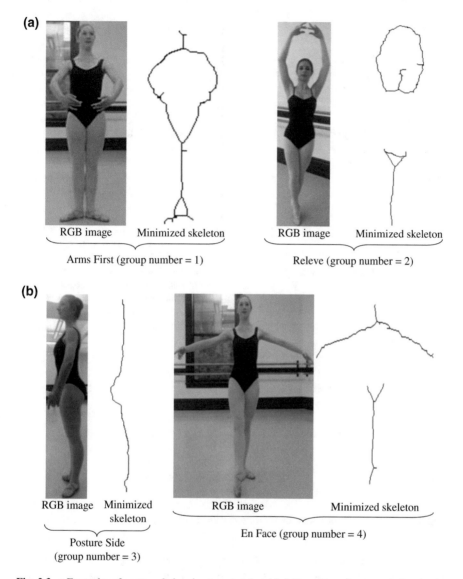

**Fig. 2.3** **a** Examples of postures belonging to group 1 and 2. **b** Examples of postures belonging to group 3 and 4

points added at regular intervals simply copy the existing value of its preceding point. The mathematical formulations shown below represent the scaling operation.

$$\lambda(l) = \begin{array}{l} 50 - (l(\mathrm{mod}\ 50))\ \textit{if}\ l(\mathrm{mod}\ 50) > 25 \\ l(\mathrm{mod}\ 50)\ \textit{if}\ l(\mathrm{mod}\ 50) \leq 25 \end{array} \qquad (2.4)$$

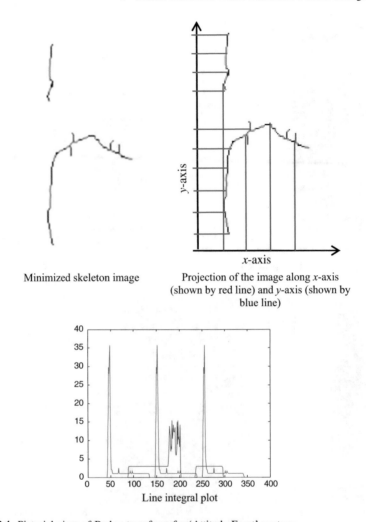

Fig. 2.4 Pictorial view of Radon transform for 'Attitude Front' posture

where $l$ refers to the number of points in the line integral plot except the ones corresponding to the leading and trailing zeroes. $\lambda$ refers to the number of points that need to be added or subtracted from the existing line integral plot.

As the function states, the number of points that needs to be added or subtracted depends on the $l$ (mod 50) value. If the value is <=25, then that number of points are deleted from the line integral plot. The points are chosen at regular interval and in general do not alter the plot to any considerable extent. If the value is greater than 25, $50 - (l$ (mod 50)) points are added to the plot. These points are added at regular intervals and it replicates the value of its preceding point. After scaling, 50 points each sampled at regular intervals from both the axis are considered for both the unknown and known postures. The sampling frequency *samp_rate* is given as

$$samp\_rate = \frac{(l + (/-)\lambda)}{50}.$$ (2.5)

Once we get 50 such points and their corresponding line integral plots with respect to both the *x*-axis and the *y*-axis, we compare the unknown and known postures. Correlation coefficients [18] corresponding to both the *x*-axis and *y*-axis are calculated for each such pair of postures. The correlation coefficient calculated along the *x*-axis is given by

$$Corr_x = \frac{\sum\limits_{m_x}(A_{m_x} - \bar{A})(B_{m_x} - \bar{B})}{\sqrt{(\sum\limits_{m_x}(A_{m_x} - \bar{A})^2)(\sum\limits_{m_x}(B_{m_x} - \bar{B})^2)}}.$$ (2.6)

While the correlation coefficient calculated along the *y*-axis is given by

$$Corr_y = \frac{\sum\limits_{m_y}(A_{m_y} - \bar{A})(B_{m_y} - \bar{B})}{\sqrt{(\sum\limits_{m_y}(A_{m_y} - \bar{A})^2)(\sum\limits_{m_y}(B_{m_y} - \bar{B})^2)}}.$$ (2.7)

Here *m* varies from 1 to 50 for both (2.12) and (2.13) and the correlation coefficient is calculated between the sample points present in the *x* and *y*-axis corresponding to the unknown posture and one of the known postures belonging to the same group as suggested by the Euler number of the unknown posture.

If the summation of the correlation coefficients is greater than a pre-determined threshold, the unknown posture is classified as the primitive posture having the highest sum of correlation coefficient value. The summation of the correlation coefficients is given by

$$Corr_{sum} = Corr_x + Corr_y.$$ (2.8)

It must be greater than the empirically determined threshold $\tau$ to qualify as a valid identified posture. The threshold value ($\tau$) for the proposed algorithm is 1.0 and is calculated empirically. The unknown posture is identified as the primitive posture in the same group having maximum $Corr_{sum}$ value. Figure 2.5 corresponds to the scaling and sampling procedure where the black stars denote the new points that are added to the plot. If the sum of correlation coefficients of the unknown posture and known posture pairs is less than the pre-determined threshold, the unknown posture is classified as a Non-Dance posture.

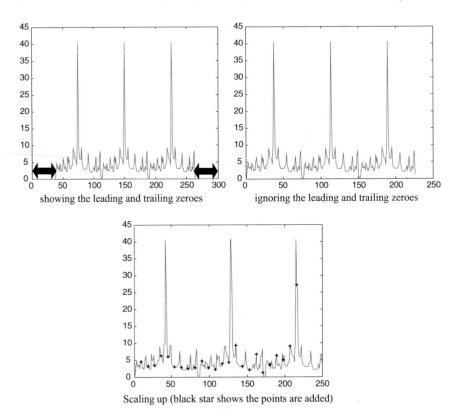

showing the leading and trailing zeroes | ignoring the leading and trailing zeroes

Scaling up (black star shows the points are added)

**Fig. 2.5** Line integral plot

## 2.3 Experimental Results

The performance of the proposed algorithm is measured in this section and it contains an exemplified detailed analysis of the classification procedure for both dance and non-dance postures.

### 2.3.1 Recognition of Unknown Posture

The Euler number of the unknown posture is 2 and it corresponds to the 4th group as shown in Table 2.2. On determination of the Euler number, Radon transform is performed on the minimized skeleton of the unknown posture. The line integral plot of the unknown posture is compared with the line integral plot of the six postures that belong to group 4. The unknown posture is recognized as the posture in the 4th group corresponding to the maximum sum of correlation coefficients. According to

**Table 2.2** Comparison between unknown posture and known 'Croise Derriere' posture

| Attributes | RGB Image | Minimum Skeleton | Euler Number | Group Number |
|---|---|---|---|---|
| (a) Unknown Posture | | | 2 | 4 |
| | Radon Transform | | | |
| | | | | |

(continued)

**Table 2.2** (continued)

| Attributes | RGB Image | Minimum Skeleton | Euler Number | Group Number |
|---|---|---|---|---|
| (b) Known 'Croise Derriere' | | | 2 | 4 |

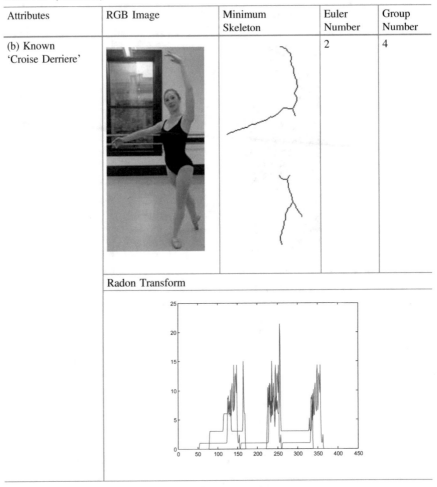

Radon Transform

the defined computations, the unknown posture is recognized as Croise Derriere which being correct, necessarily justifies the proposed algorithm. Table 2.2 presents a step by step comparison of the unknown posture and posture Croise Derriere. The initial line integral plot of the unknown posture had 539 points both along the $x$ and $y$-axes. However after ignoring the leading and trailing zeroes we arrived at 412 and 204 points along the $x$ and $y$-axes respectively. 12 and 4 points are deleted at regular intervals to scale down the plot respectively. Finally 50 points are sampled along the $x$ and $y$-axes respectively. The same sets of plots are illustrated for posture Croise Derriere with which the unknown posture is recognized. The known posture had 311 and 261 points along the $x$-axis and $y$-axis respectively. In a similar way 50 points are sampled in the $x$ and $y$-axes respectively. Once equal set of points and

**Table 2.3** Comparison between unknown posture and postures belonging to group 4

| | Known Postures | | | | | |
|---|---|---|---|---|---|---|
| | Names | $Corr_x$ | $Corr_y$ | $Corr_{sum}$ | $>\tau$ ? | Index of Best Matched Posture |
| Unknown Posture | Attitude Front | −0.10724 | −0.4638 | −0.57104 | No | 9 |
| | Croise Derriere | 0.82216 | 0.61589 | 1.43805 | Yes | |
| | Croise Devant | 0.47912 | 0.53202 | 1.0111 | Yes | |
| | Ecarte Devant | −0.21399 | 0.022809 | −0.19118 | No | |
| | Efface Devant | 0.36038 | 0.37146 | 0.73184 | No | |
| | En Face | −0.13536 | −0.23761 | −0.37297 | No | |

their values are determined, the sum of correlation coefficient is determined. Table 2.3 presents the correlation coefficients corresponding to the $x$ and $y$ axis of the sampled plot of the unknown posture and the known postures present in group number 4. Now here $Corr_{sum}$ is greater than $\tau$ for more than one posture, so the posture with highest sum of correlation coefficient value is taken as the result. The $Corr_{sum}$ values of unknown posture when comparing with Croise Derriere and Croise Devant is obtained as 1.43805 and 1.0111 respectively. As 1.43805 is greater than 1.0111, thus the result is posture Croise Derriere with $Corr_{sum}$ as 1.43805. This value is way above the threshold and is the maximum value corresponding to group number 4 for unknown posture.

An unknown posture when fails to register a sum of correlation coefficient value greater than 1.0 with any posture belonging to its group, is recognized as a Non-Dance posture. Table 2.4 depicts a non-dance posture using RGB image, minimized skeleton, line integral plot and also scaled and sampled plot. The sampled plot is compared with sampled plots of the postures present in group 5 (the Euler number of the unknown posture is 3). Table 2.4 also contains the sum of correlation values calculated with respect to the unknown posture and all the postures present in group 5 and none of them is above the predetermined threshold $\tau = 1.0$. Hence, the unknown posture is classified as a non-dance posture. Thus the proposed algorithm successfully tackles the problem of false acceptance.

## 2.3.2 Determining Correctness of a Performed Posture

The same algorithm is used to determine the correctness of a particular posture. Here, determination of the group with the help of Euler number is irrelevant as the

**Table 2.4** Comparison between unknown posture and postures belonging to group 5

| Unknown Posture | |
|---|---|
| RGB image of unknown posture | Minimized skeleton |
| | |
| Line integral plot | Sampled line integral plot |
| | |

| Known Postures | | | | |
|---|---|---|---|---|
| Names | $Corr_x$ | $Corr_y$ | $Corr_{sum}$ | $> \tau$ ? |
| Arabesque | −0.0271 | 0.271 | 0.244 | No |
| Arms Second | −0.0122 | 0.259 | 0.247 | No |
| Attitude Back | −0.0542 | −0.028 | −0.082 | No |
| Attitude Side | −0.0605 | 0.307 | 0.246 | No |

posture with which the unknown posture is supposed to be matched is known a priori. The line integral plots of both the unknown posture and Posture Front are considered. They are scaled and sampled in the same way as discussed earlier. Table 2.5 presents a step by step comparison of the unknown posture and posture Posture Front. The correlation coefficient of the unknown posture calculated along the x-axis with posture front is 0.76553 and that calculated along the y-axis is

**Table 2.5** Comparison between unknown posture and known 'Posture Front'

| Attributes | RGB Image | Minimum Skeleton | Euler Number | Group Number |
|---|---|---|---|---|
| (a) Unknown Posture | | | −1 | 2 |
| | Radon Transform | | | |
| | | | | |

(continued)

**Table 2.5** (continued)

| Attributes | RGB Image | Minimum Skeleton | Euler Number | Group Number |
|---|---|---|---|---|
| (b) Known 'Posture Front' | | | −1 | 2 |

| Radon Transform |
|---|
| |

0.26026. The summation of the correlation coefficient values is 1.0258 being greater than the threshold $\tau = 1.0$ is adjudged as a correct posture.

### 2.3.3 Determining Unknown Postures from Video Sequence

The proposed work is mainly dedicated to static 2D posture recognition from images, but can successfully be implemented in posture recognition from video sequences [19]. We have acquired a ballet video from popular website you tube (https://www.youtube.com/watch?v=b3bawTEPLtA) and break it into frames using Matlab R2013b. Now whenever we have to recognize a posture from a specific frame, we need to capture the whole body gesture, not a part of it. The video with which we are dealing with contains several frames where the whole body of the dancer is not visible, thus we have neglected those frames. And also it is not feasible to provide results for all the 900 frames present in the video, thus we are giving results only for a few frames in Table 2.6.

### 2.3.4  Performance Analysis

The performance metrics include Precision, Recall, Accuracy and F1_Score. If the True Positive, True Negative, False Positive and False Negative samples are denoted by *TP*, *TN*, *FP* and *FN* respectively.

$$Precision = \frac{TP}{TP + FP} \tag{2.9}$$

$$Recall = \frac{TP}{TP + FN} \tag{2.10}$$

$$Accuracy = \frac{TP + TN}{TP + TN + FP + FN} \tag{2.11}$$

$$F1\_Score = 2\frac{Precision \times Recall}{Precision + Recall} \tag{2.12}$$

$$Recognition\ rate = Accuracy \times 100\% \tag{2.13}$$

The overall performance of the proposed algorithm is presented in Table 2.7. The recognition rate is high when a group contains less no of primitive postures with dissimilar Radon transforms. Group 1 consists of only 2 postures, but the Radon transform of the two postures are nearly similar, so the true recognition rate falls down. But group 2 and 3 are composed of three and two postures (respectively with different Radon transforms. Thus the recognition accuracy goes up for such instances. Three postures belong to group 5 with distinctly different Radon transforms, so the recognition accuracy again goes up for all such instances. The overall true recognition rate is 91.35% where 190 out of 208 unknown postures are classified correctly. The algorithm registers recognition accuracy of 80.16, 100, 86.11,

**Table 2.6** Results obtained from an unknown video sequence

| Frame Number = 341 |
|---|

| RGB Image |  |
|---|---|

Posture is recognized as 'En Face'

| Frame Number = 351 |
|---|

| RGB Image |  |
|---|---|

Posture is recognized as 'En Face'

| Frame Number = 362 |
|---|

| RGB Image |  |
|---|---|

Posture is recognized as 'En Face'

**Table 2.7** Performance evaluation of proposed algorithm based on each class

| Posture Name | No. of Postures Taken | No. of Correctly Identified Posture | Individual Recognition Rate | Group-wise Recognition Rate |
|---|---|---|---|---|
| Arms First | 7 | 5 | 71.4286 | 80.1587 |
| Posture Front | 9 | 8 | 88.8889 | |
| Arms Third | 9 | 9 | 100.0000 | 100.0000 |
| Arms Fourth | 10 | 10 | 100.0000 | |
| Releve | 7 | 7 | 100.0000 | |
| Arms Fifth | 9 | 8 | 88.8889 | 86.1111 |
| Posture Side | 6 | 5 | 83.3333 | |
| Attitude Front | 6 | 6 | 100.0000 | 89.3386 |
| Croise Derriere | 5 | 4 | 80.0000 | |
| Croise Devant | 4 | 4 | 100.0000 | |
| Ecarte Derriere | 4 | 3 | 75.0000 | |
| EcarteDevant | 6 | 5 | 83.3333 | |
| Efface Derriere | 5 | 4 | 80.0000 | |
| Efface Devant | 7 | 6 | 85.7143 | |
| En Face | 8 | 8 | 100.0000 | |
| Epaule | 4 | 4 | 100.0000 | |
| Arabesque | 6 | 6 | 100.0000 | 94.9495 |
| Arms Second | 9 | 8 | 88.8889 | |
| Attitude Back | 9 | 9 | 100.0000 | |
| Attitude Side | 11 | 10 | 90.9091 | |
| Non-dance Posture | 67 | 61 | 91.0448 | 91.0448 |
| Overall | 208 | 190 | 91.3462 | |

89.34, 94.95 and 91.04% for unknown postures belonging to Group 1, Group 2, Group 3, Group 4, Group 5 and Non-Dance postures respectively. The overall confusion matrix is formed using $TP = 0.9513$, $FN = 0.0478$, $FP = 0.1225$ and $TN = 0.8775$.

Here we have extracted mainly two types of features Euler number and line integral plot (after ignoring leading and trailing zeros and applying sampling) from each image. These two features are combined together to form the feature space. This is the backbone of our proposed work. But if only Radon transform is applied on the minimized skeleton images (i.e., if we do not group the postures based on Euler number and without making any modifications on the line integral plots like discarding leading and trailing zeros), then the line integral plots generated can also be treated as features. And these features can be used to examine the performance of the proposed work with a comparative framework which includes Support Vector Machine (SVM) classifier [20], k-Nearest Neighbour (kNN) classification [21], Levenberg–Marquardt Algorithm induced Neural Network (LMA-NN) [22], fuzzy C means clustering (FCM) [23], Ensemble Decision Tree (EDT) [24], algorithms in [9, 10]. The parameters of all the other competing classifiers are tuned by noting the best performances after experimental trials. SVM has been used with a Radial Basis Function kernel whose kernel parameter has a value 1 and the classifier is tuned with a cost value of 100.

The performance of kNN has been reported for $k = 5$ using Euclidean distance as the similarity measure and Majority Voting to determine the class of the test samples. For LMA-NN the number of neurons in the intermediate layer is taken as 10, the value of the blending factor between gradient descent and quadratic learning as 0.01, the increase and decrease factors of the blending factor as 10 and 0.1 respectively and the stopping condition is taken as the attainment of minimum error gradient value of 1e-6. Ensemble Decision Tree classifier is used based on the principle of Adaptive Boosting taking maximum iterations as 100. The average computation time for each image is calculated in Intel Core i3 Processor with non-optimized Matlab R2013b implementation. The values of the performance metrics of the algorithms are presented in Table 2.8. It is evident from Table 2.8 that our proposed algorithm provides best results in all parametric measures stated in (2.9–2.13) and also in computation time.

A comparative evaluation of the proposed procedure is difficult due to the non-availability of a standardized database. But here we have given an assessment of the proposed work with several existing works widely varying from dance to healthcare application areas.

The proposed algorithm uses Euler number and Radon transform for feature extraction from skeleton images of dance postures. There is another way of extracting features from skeleton images using shape matching techniques. These shape detection procedures [25] are based on mainly curvature [26], similarity measures based on bottle neck distance metric [27], turning function distance [28]. The accuracy obtained using these processes drop to 82.61, 86.35 and 79.27% respectively. This degradation is due to the imperfections present in the skeleton images. To elaborate it, we are taking the minimized skeletons from Table 2.2. In Fig. 2.6, the zooming of a particular portion of the minimized skeleton images indicates the difference between the skeletons from each other when they are broken into pixel form. Thus it is evident that Radon transform is the best choice for posture recognition in ballet dance.

The algorithm finds a concrete place for posture recognition for single dancer performing ballet. But the proposed work cannot be implemented for more than one dancer as it is not possible to identify the dancers separately. Figure 2.7 explains

**Table 2.8** Comparison of performance parameters with existing literatures

| Parameters | Proposed algorithm | Algori-thm in [9] | Algor-ithm in [10] | SVM | kNN | LMA-NN | FCM | EDT |
|---|---|---|---|---|---|---|---|---|
| Recall | 0.9513 | 0.8922 | 0.7390 | 0.8488 | 0.8790 | 0.8148 | 0.7951 | 0.8382 |
| Precision | 0.8859 | 0.7780 | 0.7766 | 0.9164 | 0.8660 | 0.7506 | 0.8112 | 0.7839 |
| Accuracy | 0.9135 | 0.8663 | 0.7619 | 0.7903 | 0.8696 | 0.8941 | 0.7349 | 0.7239 |
| F1_score | 0.9471 | 0.9189 | 0.8789 | 0.8577 | 0.9143 | 0.8931 | 0.8900 | 0.8075 |
| Computation time (ms) | 2.6527 | 2.7264 | 2.4567 | 3.2951 | 2.9874 | 10.3644 | 3.1601 | 16.0215 |

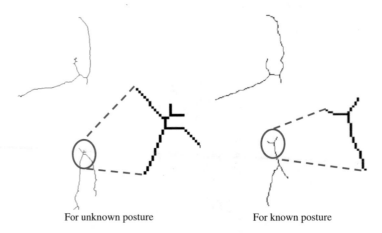

For unknown posture                              For known posture

**Fig. 2.6** Comparison of skeleton images for two Croise Derriere postures from Table 2.2

two types of RGB images where more than one person is dancing, the main problem here is that we cannot relate the upper body of the dancer to its lower body in the skeleton images Fig. 2.7a, b (ii). If we can overcome this difficulty, then the proposed algorithm can be applied to recognize ballet dance postures where multiple persons are dancing.

### 2.3.5   McNemar's Statistical Test

Let $f_A$ and $f_B$ be two classification algorithms, both the algorithms having a common training set $R$. Let $n_{01}$ be the number of examples misclassified by $f_A$ but not by $f_B$, and $n_{10}$ be the number of examples misclassified by $f_B$ but not by $f_A$. Then under the null hypothesis that both algorithms have the same error rate, the McNemar's statistic $Z$ follows a $\chi^2$ distribution with a degree of freedom equal to 1 [29].

$$Z = \frac{(|n_{01} - n_{10}| - 1)^2}{n_{01} + n_{10}} \tag{2.14}$$

Let $f_A$ be the proposed algorithm and $f_B$ be one of the other five algorithms. In Table 2.9, the null hypothesis has been rejected, if $Z > 3.84$, where 3.84 is the critical value of the Chi square distribution for 1 degree of freedom at probability of 0.05.

**Fig. 2.7  a** Example
1—Results obtained when
more than one dancer is
performing. **b** Example
2—Results obtained when
more than one dancer is
performing

**(a)**

RGB image

Minimize skeleton for respective image

**(b)**

RGB image

Minimize skeleton for respective image

**Table 2.9** Performance analysis using McNemar's test

| Competitor Algorithm $= f_B$ | Control Algorithm $f_A$ = Proposed algorithm | | | |
|---|---|---|---|---|
| | $n_{01}$ | $n_{10}$ | $Z$ | Comment |
| Algorithm in [9] | 1 | 16 | 11.5294 | Reject |
| Algorithm in [10] | 3 | 18 | 9.3333 | Reject |
| SVM | 12 | 22 | 2.3824 | Accept |
| kNN | 8 | 20 | 4.3214 | Reject |
| LMA-NN | 6 | 34 | 18.2250 | Reject |
| FCM | 4 | 42 | 29.7609 | Reject |
| EDT | 3 | 43 | 33.0652 | Reject |

## 2.4   Conclusion

Not much work has been done in the field of posture recognition of ballet. The proposed system deals with twenty fundamental dance primitives and registers an overall recognition rate of 91.35%. The procedure stated here uniquely deals with a wide range of critical dance postures. The entire endeavor proves cost effective as a single static camera produces the necessary input images for the proposed algorithm. The proposed algorithm is independent of the body type, height and weight of the ballet dancer, and hence provides even more flexibility to the learning process. The input images need not be centralized as well. The algorithm simultaneously addresses the problem of posture recognition and determination of correctness of a particular posture, thereby enhancing the effectiveness of the proposed procedure. Considering the complexity of the dance postures, an average computation time of 2.6527 s in an Intel Core i3 processor running Matlab R2013b is highly effective when compared with other standard pattern recognition algorithms.

However, certain shortcomings still do exist. The major pre-processing for this proposed work involves skin color segmentation. The dress of the ballet dancer and the background in which s/he is performing needs to be selected carefully. So to efficiently use this algorithm in other dance forms, we need to keep in mind this segmentation procedure. The proposed algorithm is not rotation invariant; therefore the danseuse must perform in a plane which is nearly parallel to the axis of the camera. Moreover the performance of the algorithm drops in case of nearly identical primitives, and the proposed algorithm in some cases fails to differentiate between them. These insufficiencies provide us with a lot of scope for further improvement over the proposed algorithm.

In a nutshell, the system proposed for posture recognition of ballet dance may be considered as a relatively unexplored application area, and the proposed system is an attempt to address the problem with reasonable accuracy with lot of scopes for future research.

## 2.5 Matlab Codes

```
% Code is written by Sriparna Saha
% Under the guidance of Prof. Amit Konar
I=imread('croiseDevant_1.jpg');
I=double(I);
I=imresize(I,0.5);
figure, imshow(I)
title('ORIGINAL IMAGE')
[hue,s,v]=rgb2hsv(I);
%%%%%%%%%% SKIN COLOR SEGMENTATION %%%%%%%%%%%
cb = 0.148* I(:,:,1) - 0.291* I(:,:,2) + 0.439 * I(:,:,3) + 128;
cr = 0.439 * I(:,:,1) - 0.368 * I(:,:,2) -0.071 * I(:,:,3) + 128;
[w h]=size(I(:,:,1));
for i=1:w
  for j=1:h
    if 140<=cr(i,j) & cr(i,j)<=165 & 140<=cb(i,j) & cb(i,j)
<=195 & 0.01<=hue(i,j) & hue(i,j)<=0.1
        segment(i,j)=1;
    else
        segment(i,j)=0;
    end
  end
end
im(:,:,1)=I(:,:,1).*segment;
im(:,:,2)=I(:,:,2).*segment;
im(:,:,3)=I(:,:,3).*segment;
figure, imshow(uint8(im))
title('SKIN COLOR IMAGE')
figure, imshow(~segment)
title('SKIN COLOR SEGMENTED IMAGE')
%%%%%%%%%% MORPHOLOGICAL OPERATIONS %%%%%%%%%%
BW1 = bwmorph(segment,'dilate',1);
figure, imshow(~BW1)
title('DILATED IMAGE')
BW2 = bwmorph(BW1,'skel',Inf);
figure, imshow(~BW2)
title('SKELETONIZED IMAGE')
BW3 = bwmorph(BW2,'spur',16);
figure, imshow(~BW3)
title('SPURRED IMAGE')
```

**Results**: Sample run results are given in Fig. 2.2.

```
BW3=~I;
 %%%%%%%%%% RADON  TRANSFORMATION %%%%%%%%%%%%
R1 = radon(BW3,0);
plot(R1(:,1),'r')
hold on
R2 = radon(BW3,90);
plot(R2(:,1),'b')
```

**Results**: Sample run results are given in Fig. 2.4.

```
load R1.mat
a=R1;
plot(a,'r')
title('With leading and trailing zeros')
al=length(a);
a1=[];%take the index of the points whose value is zero
for i=1:al
   if a(i)==0
     a1=[a1,i];
   end
end
a11=length(a1);
a2=[];%posture not centered
for i=1:(a11-1)
   if (a1(i+1)-a1(i))==1
     a2=[a2,a1(i)];
   else
     a2=[a2,a1(i)];
     break
   end
end
for i=a11:-1:1
   if (a1(i)-a1(i-1))==1
     a2=[a2,a1(i)];
   else
     a2=[a2,a1(i)];
     break
   end
end
a21=length(a2);
```

```matlab
for i=1:(a21-1)
   if (a2(i+1)-a2(i)) ~=1
      flg1_a=a2(i);
      break
   end
end
flg2_a=a2(a21);
a_req=[];
for i=(flg1_a+1):(flg2_a-1)
   a_req=[a_req;a(i)];
end
figure, plot(a_req,'r')
title('Without leading and trailing zeros')
la_req=length(a_req);
flag=mod(la_req,50);
if flag>25
   flag1=(la_req)+50-flag;
   a_req1=zeros(flag1,1);
   flag2=floor(la_req/(50-flag));
   counter=0;
   t=1;
   for i=1:flag1
      a_req1(i,1)=a_req(t,1);
      counter=counter+1;
      if counter==(flag2-1)
         a_req1(i+1,1)=a_req1(i,1);
         i=i+1;
         counter=0;
      else
         t=t+1;
      end
   end
else
   flag1=(la_req)-flag;
   a_req1=zeros(flag1,1);
   counter=0;
   t=1;
   flag2=floor(la_req/flag);
   for i=1:la_req
      if counter==(flag2-1)
         counter=0;
      else
         a_req1(t,1)=a_req(i,1);
         t=t+1;
```

```
            counter=counter+1;
        end
    end
end
figure, plot(a_req1,'r')
hold on
la_req1=length(a_req1);
counter=0;
for i=1:la_req1
    if mod(i,flag2)==0
        counter=counter+1;
        plot(i,a_req1(i),'black*')
        if counter==(50-flag)
            break
        end
    end
end
title('After adding/removing points')
samp_len=(length(a_req1))/50;
t=1;
sum=0;
for i=1:(length(a_req1))
    %sum=sum+a_req1(i,1);
    if mod(i,samp_len)==0
        a_req2(t,1)=a_req1(i,1);%sum/samp_len;
        t=t+1;
        sum=0;
    end
end
figure, plot(a_req2,'r')
title('Sampled plot')
```

**Results**: Sample run results are given in Fig. 2.5.

```
%%%%% Correlation
matchx=corr2(a_req2,c_req2);
% a and c are the sampled Radon transform plots for known and unknown postures
respectively
disp(['Correlation coefficient along x axis is ',num2str(matchx)])
matchy=corr2(b_req2,d_req2);
% b and d are the sampled Radon transform plots for known and unknown postures
respectively
disp(['Correlation coefficient along y axis is ',num2str(matchy)])
```

```
disp(' ')
sum_corr=matchx+matchy;
disp(['Summation of Correlation coefficients is ,num2str(sum_corr)])
```

**Results**: Sample run results are given in Table 2.3.

# References

1. A. LaViers, Y. Chen, C. Belta, M. Egerstedt, Automatic sequencing of ballet poses. IEEE Robot. Autom. Mag. **18**(3), 87–95 (2011)
2. F. Guo, G. Qian, Dance posture recognition using wide-baseline orthogonal stereo cameras, in *7th International Conference on Automatic Face and Gesture Recognition, FGR 2006* (2006), pp. 481–486
3. P. Silapasuphakornwong, S. Phimoltares, C. Lursinsap, A. Hansuebsai, Posture recognition invariant to background, cloth textures, body size, and camera distance using morphological geometry, in *2010 International Conference on Machine Learning and Cybernetics (ICMLC)*, vol 3 (2010), pp. 1130–1135
4. K. Akita, Image sequence analysis of real world human motion. Pattern Recognit. **17**(1), 73–83 (1984)
5. Y. Guo, G. Xu, S. Tsuji, Understanding human motion patterns, in *Proceedings of the 12th IAPR International. Conference on Pattern Recognition, 1994. Vol. 2-Conference B: Computer Vision & Image Processing*, vol 2 (1994), pp. 325–329
6. C. Hu, Q. Yu, Y. Li, S. Ma, Extraction of parametric human model for posture recognition using genetic algorithm, in *Proceedings of the 4th IEEE International Conference on Automatic Face and Gesture Recognition* (2000), pp. 518–523
7. G. Feng, Q. Lin, Design of elder alarm system based on body posture reorganization, in *2010 International Conference on Anti-Counterfeiting Security and Identification in Communication (ASID)* (2010), pp. 249–252
8. Z. Zhou, W. Dai, J. Eggert, J.T. Keller, M. Rantz, Z. He, A real-time system for in-home activity monitoring of elders, in *Annual International Conference of the IEEE Engineering in Medicine and Biology Society, EMBC 2009* (2009), pp. 6115–6118
9. S. Saha, A. Banerjee, S. Basu, A. Konar, A.K. Nagar, Fuzzy image matching for posture recognition in ballet dance, in *2013 IEEE International Conference on Fuzzy Systems (FUZZ)* (2013), pp. 1–8
10. A. Banerjee, S. Saha, S. Basu, A. Konar, R. Janarthanan, A novel approach to posture recognition of ballet dance, in *2014 IEEE International Conference on Electronics, Computing and Communication Technologies (IEEE CONECCT)* (2014), pp. 1–5
11. S. Saha, S. Ghosh, A. Konar, A.K. Nagar, Gesture recognition from indian classical dance using kinect sensor, in *2013 5th International Conference on Computational Intelligence, Communication Systems and Networks (CICSyN)* (2013), pp. 3–8
12. S. Saha, S. Ghosh, A. Konar, R. Janarthanan, Identification of Odissi dance video using Kinect sensor, in *2013 International Conference on Advances in Computing, Communications and Informatics (ICACCI)* (2013), pp. 1837–1842
13. S.L. Phung, A. Bouzerdoum Sr., D. Chai Sr., Skin segmentation using color pixel classification: analysis and comparison. IEEE Trans. Pattern Anal. Mach. Intell. **27**(1), 148–154 (2005)
14. M.J. Jones, J.M. Rehg, Statistical color models with application to skin detection. Int. J. Comput. Vis. **46**(1), 81–96 (2002)

15. C.R. Dyer, Computing the Euler number of an image from its quadtree. Comput. Graph. Image Process. **13**(3), 270–276 (1980)
16. S.R. Deans, Hough transform from the Radon transform. IEEE Trans. Pattern Anal. Mach. Intell. **2**, 185–188 (1981)
17. G. Beylkin, Discrete radon transform. IEEE Trans. Acoust. Speech Signal Process. **35**(2), 162–172 (1987)
18. J.P. Lewis, Fast normalized cross-correlation. Vision Interface **10**(1), 120–123 (1995)
19. A. Chaudhary, J.L. Raheja, Bent fingers' angle calculation using supervised ANN to control electro-mechanical robotic hand. Comput. Electr. Eng. **39**(2), 560–570 (2013)
20. S. Theodoridis, A. Pikrakis, K. Koutroumbas, D. Cavouras, *Introduction to Pattern Recognition: A Matlab Approach: A Matlab Approach* (Academic Press, 2010)
21. T.M. Mitchell, Machine learning and data mining. Commun. ACM **42**(11), 30–36 (1999)
22. S. Saha, M. Pal, A. Konar, R. Janarthanan, Neural network based gesture recognition for elderly health care using kinect sensor, in *Swarm, Evolutionary, and Memetic Computing* (Springer, 2013), pp. 376–386
23. M. Pal, S. Saha, A. Konar, A fuzzy C means clustering approach for gesture recognition in healthcare. Knee **1**, C7
24. T.G. Dietterich, An experimental comparison of three methods for constructing ensembles of decision trees: bagging, boosting, and randomization. Mach. Learn. **40**(2), 139–157 (2000)
25. R.C. Veltkamp, Shape matching: similarity measures and algorithms, in *SMI 2001 International Conference on Shape Modeling and Applications* (2001), pp. 188–197
26. F. Mokhtarian, S. Abbasi, Retrieval of similar shapes under affine transform, in *Proceedings of the 3rd International Conference on Visual Information and Information Systems* (1999), pp. 566–574
27. A. Efrat, A. Itai, Improvements on bottleneck matching and related problems using geometry, in *Proceedings of the 12th Annual Symposium on Computational Geometry* (1996), pp. 301–310
28. E.M. Arkin, L.P. Chew, D.P. Huttenlocher, K. Kedem, J.S.B. Mitchell, An efficiently computable metric for comparing polygonal shapes. IEEE Trans. Pattern Anal. Mach. Intell. **13**(3), 209–216 (1991)
29. T.G. Dietterich, Approximate statistical tests for comparing supervised classification learning algorithms. Neural Comput. **10**(7), 1895–1923 (1998)

# Chapter 3
# Fuzzy Image Matching Based Posture Recognition in Ballet Dance

This chapter provides a fuzzy logic based approach to posture recognition in Ballet dance. The initial section of the chapter deals with a brief introduction to fuzzy sets and its characteristics, while the later part of the chapter is entirely devoted to the discussion of the technique. The techniques of chain code generation, straight line approximation and matching using a fuzzy T-norm operator populate the contents of the chapter thereby illustrating and validating the suggested algorithm.

## 3.1 Introduction

Ballet is an artistic dance form, performed to music using unequivocal and highly formalized set steps and kinesics. The dance form originated in Italy in the fifteenth century. The present form of the dance is a result of cultural transfusion, especially from countries like France and Russia. Ballet consists of 17 static postures, namely arabesque, arms first, arms second, arms third, arms fourth, arms fifth, attitude front, attitude back, attitude side, croise derriere, croise devant, ecarte devant, efface devant, en face, posture front, posture back, and releve.

The proposed body of work deals with fuzzy posture recognition of the various dance sequences of Ballet. Recognition, modelling and analysis of the various postures comprising the dance form help an enthusiast to learn the art by intelligibly interacting with the proposed system. The danseuse interacts with the computer and in the process learns the dance form in a simple and flexible manner.

While dealing with identification of Ballet postures, the initial problem that arises is segmentation of the dancer from the background. Ballet is performed adhering to a specific dress code, by virtue of which the head, hands and legs of the dancer are distinctly visible. Skin color segmentation is performed on the input

Contributed by Anupam Banerjee, Sriparna Saha and Amit Konar

© Springer International Publishing AG 2018
A. Konar and S. Saha, *Gesture Recognition*, Studies in Computational
Intelligence 724, DOI 10.1007/978-3-319-62212-5_3

images, to get rid of the unnecessary background information. The segmentation procedure extracts the Ballet dancer from the background thereby enabling us to concentrate on the posture of the danseuse. The input images belong to the RGB color space thereby inherently succumbing to the pitfalls of high correlation, non-uniformity and mixing of chrominance and luminance. The segmentation procedure as described in Sect. 2.2.1. With the help of this procedure, only skin color is detected and unwanted information pertaining to the background is neglected. On performing skin color segmentation [1], the authors skeletonize [2, 3] the segmented structure with the help of morphological operations [4, 5]. In the next step a fixed number of pixels are spurred from the resultant skeleton, which successfully minimizes the skeleton by pruning insignificant structures from it. Representing postures with the help of minimized skeletons reduces the influence of body structure and weight of the Ballet dancers on the generated dance pattern. In the subsequent step, with the help of chain code [6] representation of unbounded regions and sampling of bounded regions, the authors perform straight line approximation on the minimized skeletons. The straight line approximated minimized skeletons correspond to stick figure diagrams of the dance postures. On successful generation of stick figure diagrams, significant straight lines that maximally represent the posture are determined. In the next step the fuzzy membership [7, 8] value of the significant lines are calculated, and by calculating their degree of belongingness to each quadrant. In the ultimate step the dominant lines are considered and their fuzzy membership values corresponding to the four quadrants to generate the initial database comprising of the seventeen fundamental primitive dance postures of Ballet. The dominant lines and their fuzzy membership values form the basis of the fuzzy T-norm [7] operator. The fuzzy membership values of unknown posture to the fuzzy membership values of the seventeen fundamental dance primitives are compared. The comparison is done with the help of the fuzzy T-norm operator, the output of which determines the proximity of the generated dance move with the seventeen fundamental Ballet dance primitives.

In this chapter, a novel technique for 2D posture recognition of Ballet dance is introduced. Sections 3.2–3.5 addresses introductory concepts of fuzzy set theory. Sections 3.6–3.14 discusses the methodologies followed in this algorithm. Section 3.15 contains a database consisting of the seventeen fundamental dance primitives of Ballet. This is followed by experimental results in Sect. 3.16. Section 3.17 concludes while Sect. 3.18 explains the Matlab codes for this chapter.

## 3.2  Fundamental Concepts

A Set is a well-defined collection of objects. The constituents of a set are its elements or members, inclusions of which are governed by the characteristics that define the set. For example, professors teaching at a particular university constitute a *PROFESSOR* set and their inclusion in the collection can be easily determined from information available in the university database. However a set named *INTELLIGENT STUDENTS* cannot be constructed unless the definition of intelligent is well defined

A set $Z$ and the belongingness of an element $b$ in it is denoted by the expression

$$b \in Z \tag{3.1}$$

Let $c$ be a member of set $Y$, such that for all $c$, $c$ is also a member of set $Z$. Then $Y$ is a subset of $Z$ and is denoted by

$$Y \subseteq Z \tag{3.2}$$

If $Y$ is subset of $Z$ where every element of $Z$ is not present in set $Y$ then $Y$ is a proper subset of $Z$ and is denoted by

$$Y \subset Z \tag{3.3}$$

If every element contained in $Y$ is an element of set $Z$ and vice versa then set $A$ and set $B$ are equal and is represented as

$$Y = Z \tag{3.4}$$

For any two sets $Z$ and $Y$ if there exists at least one element $x$, common to both $Z$ and $Y$, it can be said that

$$x \in (Y \cap Z) \tag{3.5}$$

where $\cap$ denotes the intersection operation.

For any two sets $Z$ and $Y$ if there exists at least one element $x$, such that $x$ belongs to either $Z$ or $Y$, it can be said that

$$x \in (Y \cup Z) \tag{3.6}$$

where $\cup$ denotes the union operation.

A universal set $U$ is domain specific and contains all possible members of that particular domain. In other words, all sets in a given domain are subsets of the universal set $U$.

## 3.3 Fuzzy Sets

The conditions defining the set boundaries in conventional sets are rigid and are discretely defined. The membership criteria adhering to a conventional set theoretic approach defines the universal set *TEMP* and its subsets named *VERY COLD*, *COLD*, *PLEASANT*, *HOT* and *VERY HOT*.

$$\text{VERY COLD } = \{\text{temp} \in \textit{TEMP}: 4\,^{\circ}\text{C} > \text{temp}\} \tag{3.7}$$

$$\text{COLD} = \{\text{temp} \in \textit{TEMP}: 4\,^{\circ}\text{C} \leq \text{temp} \leq 18\,^{\circ}\text{C}\} \tag{3.8}$$

$$\text{MODERATE} = \{\text{temp} \in \textit{TEMP}: 18\,^{\circ}\text{C} \leq \text{temp} \leq 30\,^{\circ}\text{C}\} \tag{3.9}$$

$$\text{HOT} = \{\text{temp} \in \textit{TEMP}: 30\,^{\circ}\text{C} \leq \text{temp} \leq 40\,^{\circ}\text{C}\} \tag{3.10}$$

$$\text{VERY HOT} = \{\text{temp} \in \textit{TEMP}: 40\,^{\circ}\text{C} < \text{temp}\} \tag{3.11}$$

The definition of the sets validates the discreteness of each set. A temperature value of 29.99 °C will belong to set *MODERATE* however an increase in just 0.01 °C will make it a member of set *HOT*. Such discrete behaviour fails to capture the essence of imprecise real world systems. Such sharp demarcation of set boundaries fails to model non discrete systems and setups.

In conventional set theory, members are either assigned membership value 0 or 1, based on which their belongingness in a particular set is decided. Elements having membership value 1 belong to the set whereas the ones with membership value 0 do not belong to it. The following connotations are used to describe the membership of an element b in a set Z.

$$\mu_b(Z) = 1 \tag{3.12}$$

$$\mu_b(Z) = 0 \tag{3.13}$$

Fuzzy sets follow a different principle, by virtue of which it assigns a spectrum of membership values in the interval [0,1] to its elements. Additionally every element of the universal set is a member of a given set Z. Thus for every element $b \in U$,

$$0 \leq \mu_b(Z) \leq 1 \tag{3.14}$$

As all the members of the universal set $U$ belong to fuzzy set Z, the boundary definitions of two fuzzy sets Z and X may overlap. For example, in contrast to the respective conventional sets: *VERY COLD, COLD, MODERATE, HOT* and *VERY HOT*, the corresponding fuzzy sets allow any temperature as a member of each of the above sets but with different membership values in the range [0, 1]. As a specific instance the temperature 20 °C is a member of all the fuzzy sets but with different membership values of 0.001, 0.1, 1.00, 0.60 and 0.20 corresponding to the sets *VERY COLD, COLD, MODERATE, HOT* and *VERY HOT* respectively. The above example makes sense due to the logic that a temperature of 20 °C corresponds to a moderate temperature and therefore has highest membership value of 1.00 with respect to the set *MODERATE*. The relative grading of the other memberships can be easily comprehended from the usual meaning of the terms *VERY COLD, COLD, MODERATE, HOT* and *VERY HOT*.

A fuzzy set $Z$ is a set of ordered pairs, given by

$$Z = \{(b, \mu_z(b)) : b \in X\} \tag{3.15}$$

where $X$ is a universal set of objects (also referred to as the universe of discourse) and $\mu_Z(b)$ is the grade of membership of the object $b$ in $Z$. Usually $\mu_Z(b)$ lies in the closed interval of $[0, 1]$.

It needs to be mentioned that sometimes the range of membership is relaxed from $[0, 1]$ to $[0, R_{max}]$ where $R_{max}$ is a positive finite real number. However one can easily convert $[0, R_{max}]$ to $[0, 1]$ by dividing the membership values in the range $[0, R_{max}]$ by $R_{max}$.

Some commonly used notations to denote fuzzy sets and its membership values are as follows

$$Z = \{(b_1\mu_z(b_1)), (b_2\mu_z(b_2)), (b_3\mu_z(b_3)), \ldots, b_n\mu_z(b_n))\} \tag{3.16}$$

$$Z = \{b_1/\mu_z(b_1), b_2/\mu_z(b_2), b_3/\mu_z(b_3), \ldots, b_n/\mu_z(b_n)\} \tag{3.17}$$

$$Z = \{b_1/\mu_z(b_1) + b_2/\mu_z(b_2) + b_3/\mu_z(b_3) + \cdots + b_n/\mu_z(b_n)\} \tag{3.18}$$

$$Z = \{\mu_z(b_1)/b_1, \mu_z(b_2)/b_2, \mu_z(b_3)/b_3, \ldots, \mu_z(b_n)/b_n\} \tag{3.19}$$

$$Z = \{\mu_z(b_1)/b_1 + \mu_z(b_2)/b_2 + \mu_z(b_3)/b_3 + \cdots + \mu_z(b_n)/b_n\} \tag{3.20}$$

## 3.4 Membership Functions

The membership function $\mu_z(b)$ maps the object or its attribute b to positive real numbers in the interval $[0, 1]$. A formal definition of membership functions is provided below

A membership function $\mu_Z(b)$ is denoted by the following mapping:

$$\mu_z(b) \Rightarrow [0, 1], \quad b \in X \tag{3.21}$$

where $b$ describes an object or its attribute $X$ is the universe of discourse and $Z$ is a subset of $X$.

Consider the problem of defining *VERY COLD, COLD, MODERATE, HOT* and *VERY HOT* by their respective membership functions. Closer the temperature to 0, higher is its chance of belonging to the set *VERY COLD*. So if $b$ is the temperature, *VERY COLD* can be defined as follows:

$$VERY\ COLD = \{(b, \mu_{VERY\ COLD}(b))\} \quad where \quad \mu_{VERY\ COLD}(b) = \exp(-b) \quad (3.22)$$

As b increases the value of $\mu_{VERY\ COLD}(b)$ decreases exponentially, i.e. as $b$ approaches 0 the value of $\mu_{VERY\ COLD}(b)$ tends towards value 1. A controlled decrease can be incorporated into the existing membership function by including a factor $\alpha$. Thus,

$$\mu_{VERY\ COLD}(b) = \exp(-\alpha b)\ for\ \alpha > 0 \qquad (3.23)$$

Greater the value of $\alpha$ higher will be the falling rate of $\mu_{VERY\ COLD}(b)$ over $b$. In a similar fashion the membership values for COLD, MODERATE, HOT and VERY HOT fuzzy sets can be defined. The figure below provides a plot of the different membership functions used to represent the various sets.

The curve for the *COLD* fuzzy set has its peak at some temperature greater than 0 and has a sharp fall around the peak. The meaning of the word *COLD* logically justifies the representation of its membership function. The membership curve for the fuzzy set *MODERATE* has its peak at temperature $b = 22$ and falls off slowly on both sides around the peak. The representations of the various membership functions are subjective to a large extent and quite naturally depend on the readers' perspective. If the readers' view are different, they can allow a sharp falloff of the curve around temperature $b = 22$. One interesting point to note about the *HOT* and *VERY HOT* membership functions is that *HOT* curve throughout has a higher membership value than the *VERY HOT* curve until both of them saturate at temperature $b = 50$ °C. That is meaningful because if a temperature corresponds to the characteristics of the set *VERY HOT*, it is bound to follow the characteristics of the set *HOT*, however the converse may not always be true.

The membership functions shown in Fig. 3.1 can be represented by a variety of mathematical functions. One such representation is provided below:

$$\mu_{COLD}(b) = \frac{pb^2}{1 + qb^2 + rb}, \quad p, q, r > 0 \qquad (3.24)$$

$$\mu_{MODERATE}(b) = \exp\left[-\left(\frac{b - 24}{2\sigma^2}\right)\right], \quad \sigma > 0 \qquad (3.25)$$

$$\mu_{HOT}(b) = 1 = \exp(sb^2), \quad s > 0 \qquad (3.26)$$

$$\mu_{VERY\ HOT}(b) = 1 - (-sb), \quad s > 0 \qquad (3.27)$$

The parameters $p$, $q$, $r$ and $s$ in the membership functions defined above are selective intuitively by the experts based on their subjective judgment in the respective domains. Tuning of these parameters is needed to control the curvature and sharp changes on the curves around some selected $b$-values.

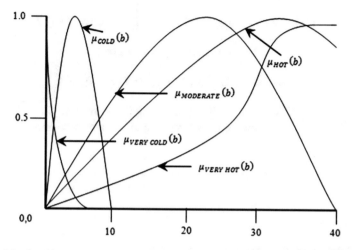

**Fig. 3.1** Membership curves for the fuzzy sets *VERY COLD, COLD, MODERATE, HOT* and *VERY HOT*. The x-axis denotes the temperature in degree Celsius and the y-axis denotes the corresponding membership values

## 3.5  Operation on Fuzzy Sets

Unlike conventional sets, the operations on fuzzy sets are usually described with reference to membership functions. Among the common operations on fuzzy sets, fuzzy T-norm, fuzzy S-norm and fuzzy complementation need special mention. In this book chapter, fuzzy T-norm operation is used. It is outlined in next subsection along with its mathematical expressions.

### 3.5.1  Fuzzy T-Norm

For any two fuzzy sets $A$ and $B$ defined with respect to a common universe of discourse $X$, the intersection of the fuzzy sets, characterized by a T-norm operator, is represented as

$$\mu_{A \cap B}(x) = T(\mu_A(x), \mu_B(x)) \tag{3.28}$$

For fuzzy membership values $a$, $b$, $c$ and $d$, the T-norm operator T can be formally defined as follows.

$$\text{Boundary: } T(0,0) = 0, \ T(a,1) = T(1,a) = a \tag{3.29}$$

$$\text{Monotonicity: } T(a,b) \leq T(c,d) \quad \text{if} \quad a \leq c \text{ and } b \leq d \qquad (3.30)$$

$$\text{Commutativity: } T(a,b) = T(b,a) \qquad (3.31)$$

$$\text{Associativity: } T(a,T(b,c)) = T(T(a,b),c) \qquad (3.32)$$

A function $T : [0,1] \, x \, [0,1] \rightarrow [0,1]$ satisfying the above four characteristics are called a T-norm. The following is the example of the typical T-norm function.

$$\text{Drastic product} \quad T_{dp}(a,b) = \begin{array}{ll} = a & \textit{if } b = 1 \\ = b & \textit{if } a = 1 \\ = 0 & \textit{otherwise} \end{array} \qquad (3.33)$$

## 3.6   Towards Algorithm Specification

The work flow diagram shown in Fig. 3.2 provides a basic framework to the proposed algorithm. Each step is explained to a considerable amount, in the following sub-Sections.

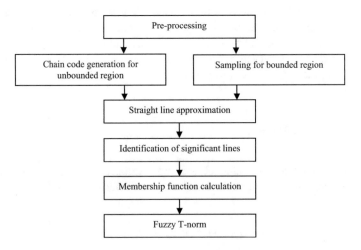

**Fig. 3.2** Block diagram of the proposed algorithm for fuzzy image matching based posture recognition in ballet dance

## 3.7  Pre-processing

The pre-processing begins with skin color segmentation of the RGB input images pertaining to the Ballet dancer. Skin color segmentation procedure is explained in details in Sect. 2.2.1. The skin color segmentation procedure successfully extract out the posture from the background [1].

After performing skin color based segmentation, skeletonization technique has been implemented on the extracted posture [2, 3]. For efficient representation of the skeleton image, shorter lines with length less than 16 pixels are removed from the skeleton using morphological spur operation. The 16 pixel threshold considered in this case is chosen empirically. The minimized skeleton finally comprises of significant lines only as the portions removed contained little or no significant information [9]. Figure 3.3 shows the minimized skeleton formation for Releve posture. These pre-processing stages for morphological operations are elaborately explained in Chap. 2.

## 3.8  Chain Code Generation for Unbounded Regions

On successful formation of minimized skeletons and removal of minor aberrations through spurring the authors proceed to the next stage of the algorithm. In this stage, chain codes corresponding to the configuration of the minimized skeletons are generated. The generated minimized skeleton is one pixel thick, except at points

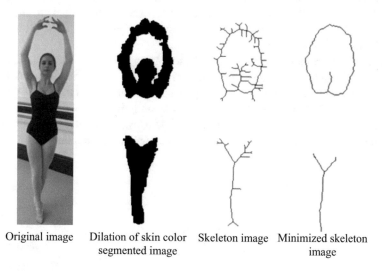

Original image    Dilation of skin color    Skeleton image    Minimized skeleton
                  segmented image                             image

**Fig. 3.3** Results obtained for 'Releve' posture during pre-processing stage

where the skeleton bifurcate to give rise to separate body parts of the Ballet dancer. For example at a bifurcation point the minimized skeleton grows in three directions, one part forms the head section, while the other two correspond to the hands of the Ballet dancer. In different instances, the minimized skeleton give rise to a number of such branch points. The minimized skeletons are encoded with the help of chain codes. At a later point of time, the stick figure diagrams of the dance postures with the help of straight line approximation regenerated, which in the proposed procedure relies heavily on the generated chain code.

The technique of chain coding was first established by Freeman [6] in and has subsequently proved to be of considerable importance in describing the contours of silhouettes [10]. Other popular chain codes include the Vertex Chain Code (VCC) [11], Three Orthogonal symbol chain code (3OT) [12] and Directional Freeman Chain Code of Eight Directions (DFCCE) [13]. The technique described by Freeman elegantly encodes arbitrary geometric configurations, which essentially facilitates their study and manipulation by means of a digital computer. A great many number of ways exists to encode arbitrary geometric structures to facilitate such manipulations, each having its own set of advantages and disadvantages. The chain coding scheme used in the   present context uses rectangular array type encoding. In this method the slope function is approximated with the help of eight standard slope measures.

In the rectangular array based chain code procedure [6], depicted in Fig. 3.4, a number sequence made up of only one kind of digit corresponds to a straight line; e.g. the sequence 22222....2 represents a straight line at an angle of 90°. All together eight such straight lines can be represented in this manner with the digits "0" through "7". Each of these straight lines makes an angle with the horizontal which is equal to the digit multiplied by 45°. Finer angle measurements and representations other than general multiples of 45° can be formed with digit pairs, triplets, or higher-order digit groups; the basic code representing the eight multiples of 45° serves our purpose.

The algorithm that is being proposed uses rectangular array based chain coding to represent minimized skeletons as shown in Fig. 3.5. In the first step the authors evaluate the number of connected components present in the minimized skeleton. Once the number of connected components and their pixel compositions are evaluated, the authors determine the end points of each connected component. On locating the end points the authors encode the connected component as a part (or

**Fig. 3.4** Freeman chain code

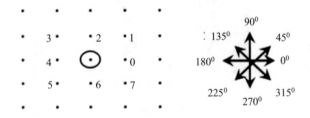

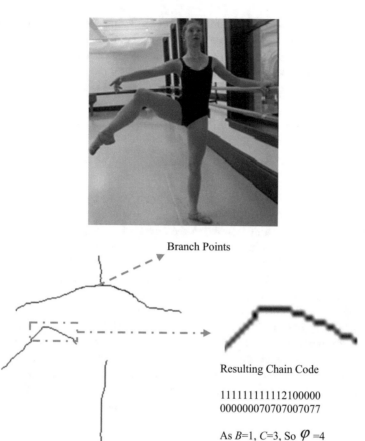

**Branch Points**

**Resulting Chain Code**

111111111112100000
000000070707007077

As $B=1$, $C=3$, So $\varphi =4$

**Fig. 3.5**  Application of chain code to 'Attitude Side' posture

sometimes in its entirety, depending on the number of branch points and connected components of the skeleton) of the minimized skeleton with the help of the previously mentioned rectangular array based chain coding. The curved structures constituting the connected components are made up of small straight lines (sometimes as small as a single pixel). The chain code is generated with the help of a sequence of digits, where each digit represents the angle encoding corresponding to a particular pixel. For example if the pixel next to the pixel under consideration is at an angle of 90° to it, the code corresponding to the pixel denoted by $C_k$ ($k$ denotes the $k$th pixel) will be equal to 2. The chain code is successfully generated except when the authors come across branch points. These bifurcation points as mentioned earlier indicate branches in the minimized skeleton. In such cases, the branch points are noted, and one of the two branches emanating out from the bifurcation point is isolated and coded separately, while the other branch is encoded in continuation with the branch conceding the bifurcation. Let us assume for a particular minimized skeleton, $C$ connected components and $B$ branch points are there, then the total number of chain codes that needs to be individually generated, corresponding to the minimized skeleton will be equal to

$$\phi = C + B \tag{3.34}$$

## 3.9  Sampling of Bounded Regions

The minimized skeletons comprises of both bounded and unbounded regions. The previous section described the treatment and encoding of unbounded regions whereas the present section focuses on the bounded contours. Encoding both type of contours need a pixel by pixel traversal of the entire connected component. While traversing the branch points are noted, and ensuring that there is no trace back, go on tracing pixels in the forward region. Once the branch point is revisited, then the trace identifies a bounded region.

The number of pixels encountered during this trace is denoted by $N_p$. The minimized skeletons consist of simple and straight forward bounded regions and keeping the computational complexity in mind, simple sampling instead of polygon approximation of the bounded regions is retorted. Depending on the value of $N_p$, the authors either sample the bounded contour to six or three equal parts. Empirical findings have validated the fact that simple sampling approximates the bounded regions to a fairly considerable extent. The sampling of the bounded regions is depicted in Fig. 3.6.

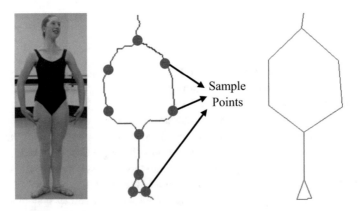

**Fig. 3.6** Application of sampling to 'Posture Front'

## 3.10   Straight Line Approximation

In this section the authors discuss the procedure adopted to represent the bounded
and unbounded regions present in the minimized skeletons with the help of straight
lines. The straight line approximation procedure adopted for bounded regions is
easy and consists of joining the sampled co-ordinate points with the help of straight
lines. The procedure adopted to approximate the unbounded region is explained in
the following part of this section.

In Sect. 3.8 the authors have discussed the encoding of unbounded regions with
the help of rectangular array based chain codes. Approximating the chain codes
with the help of straight lines give rise to stick figure diagrams corresponding to the
minimized skeletons. Keeping the length of the chain code in mind, the authors
propose a technique with the help of which the authors evaluate the points that add
maximum curvature to the minimized skeleton. Associated with each pixel is a digit
and all these digits join to form the chain code of a particular curved line segment.
The constant $S_i$ to the length of the $i$th chain code is associated, where $i$ vary from 1
to $\phi$ for a particular minimized skeleton. Now depending on the length of the chain
code $S_i$, a parameter $\alpha_i$ is determined. Equation (3.35), illustrates the formulation of
$\alpha_i$. Based on the value of $\alpha_i$, a value $\beta_n$ is evaluated for each pixel where n ranges
from $\alpha_i + 1$ to $S_i - (\alpha_i + 1)$, i.e., the value $\beta_n$ doesn't exist for the remaining pixels,
as the region under consideration is an unbounded region. $\beta_n$ is defined in (3.36)
and clearly validates the nonexistence of (3.35) for pixels numbered 1 to $\alpha_i$ and $S_i$ to
$S_i - \alpha_i$.

$$\alpha_i = f(S_i) = 10 \quad \text{for } S_i < 80$$
$$= 25 \qquad\quad \text{for } S_i \geq 80 \tag{3.35}$$

$$\beta_n = \left| \sum_{k=n}^{n+\alpha_i} C_k - \sum_{k=n}^{n-\alpha_i} C_k \right| \tag{3.36}$$

Once the value of $\beta_n$ is evaluated for each pixel corresponding to individual
chain codes, (pixels numbered 1 to $\alpha_i$ and $S_i$ to $S_i - \alpha_i$ has value of $\beta_n = 0$) the
authors proceed to the subsequent step. Here the pixels based on their values of $\beta_n$
are sorted. Once the sorting of the pixels is done the co-ordinates of the top 2 pixels
are selected having maximum value of $\beta_n$ and are at a considerable distance from
each other, and from the branch points as well (a maximum of 1 branch point is
encountered per connected component). The distance measure considered here is
empirical and depends on experimental findings.

Adhering to the mentioned protocol an undulating, unbounded, curvy portion of
the minimized skeleton is represented, corresponding to one chain code with the
help of a maximum of 4 straight lines. Smaller fragments usually tend to dissatisfy
the above mentioned conditions, and as they seldom have branch points, as a
consequence of which they are often represented with the help of a single straight

**Fig. 3.7** Straight line approximation to 'Arabesque' posture

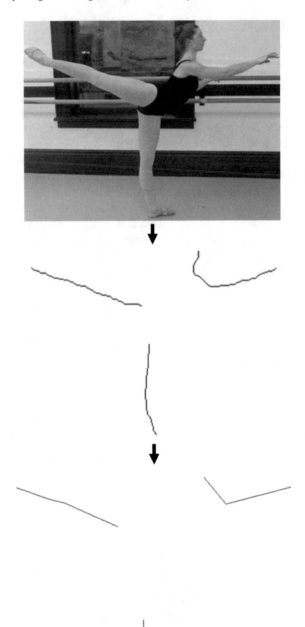

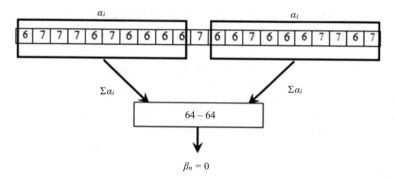

**Fig. 3.8** Procedure for performing straight line approximation

line. Once the essential co-ordinates are obtained from the above stated procedure, corresponding co-ordinates are joined to generate the stick figure diagrams for each minimized skeleton. The terseness and elegance of the chain code makes the entire procedure computationally cheap and hassle free. Chain code corresponding to pixels belonging to diametrically opposite directions differ from each other by a constant value of 4, as a result of which sharp angular variations are distinctly captured with the help of the above mentioned procedure. Special care is taken to ensure selection of pixels at a considerable distance from each other and from the branch points as well, as neighboring pixels tend to have similar $\beta_n$ values. Figure 3.7 successfully depicts the above mentioned procedure. Chain code for the extracted part is 666666666..............6666666766677767666766766667767 where the length of the segment is 75 and as $S_i < 80$ for $\alpha_i = 10$. The corresponding values of $\beta_n$ are given as: 00000000000000012.......3443201232000000000000000. Where the location corresponding to the first 4 in the sequence is the point where the line bends, and hence indicates a point of maximum curvature. Figure 3.8 corresponds to a pictorial representation of the above mentioned procedure.

## 3.11  Significant Lines Identification

The subsequent steps of the proposed algorithm demands determination of significant straight lines from the straight line approximated segments. The length of the straight lines is calculated by considering the Euclidean distance between the first and last pixel of the corresponding line segments. In order to do evaluate the significant straight lines, the lines are sorted based on their Euclidean distance, and consider the five largest straight lines for postures containing unbounded regions, and six straight lines for postures with bounded regions. Three straight lines are taken for "posture side" and "attitude back" as they are composed of 3 and 4 straight lines respectively.

## 3.12  Calculation of Fuzzy Membership Values of Significant Lines

Once the straight line approximated stick figure diagrams are generated and their corresponding significant straight lines are evaluated, their fuzzy membership values are determined. In order to determine the fuzzy membership values the typical abscissa ordinate based co-ordinate system are considered. The center co-ordinate has both its ordinate and abscissa values equal to 0. Here the fuzzy membership values represent the degree of belongingness of a straight line to the four quadrants. In a more elaborate sense, the stick figure has a certain positioning in the regular co-ordinate setup, from which the degrees of belongingness of the constituent straight lines to each quadrant are evaluated. To calculate the degree of belongingness to each quadrant, four fuzzy sets $A$, $B$, $C$ and $D$ are constructed. The straight lines that make up a stick figure diagram usually belong to a single quadrant or it spans along two adjacent quadrants. Experimental findings have asserted the fact that the lines never belong to non-adjacent quadrants, or to more than two quadrants. Let a straight line $ZX$ spans in two quadrants $C$ and $A$. Then a line $ZY$ parallel to abscissa of the image is drawn and $XY$ is the normal line on $ZY$ from $X$ co-ordinate. Further suppose the co-ordinates of $Z$ and $X$ are $(x_1, y_1)$ and $(x_2, y_2)$ respectively. Two triangles are formed $XYZ$ and $PZQ$.

Mathematically,

$$\Delta XYZ \cong \Delta PZQ \tag{3.37}$$

$$\frac{ZP}{ZX} = \frac{ZQ}{ZY} = \frac{0 - x_1}{x_2 - x_1} = \frac{x_1}{x_1 - x_2} \tag{3.38}$$

$$\frac{ZP}{ZX} = \frac{-ve}{(-ve) - (+ve)} = \frac{-ve}{-ve} = +ve \tag{3.39}$$

The fuzzy membership value or degree of belongingness of each significant straight line is evaluated with the help of (3.41–3.44). Membership value $\mu_x^{line}$ where $x = A, B, C, D$ varies between 0 to 1. The bold lines drawn below indicates the line configurations where its membership value $= 1$, whereas the dotted lines owing to its span over multiple quadrants can have its membership values greater than 0 but less than 1. The membership values of the significant lines are calculated using the understated functions. Here $(x_i, y_i)$ denotes the start and end co-ordinates of the significant straight lines. Figure 3.9 provides the basis, based on which the membership functions are calculated. Figures 3.10, 3.11, 3.12 and 3.13 presents the all possible combinations in which the constituent straight lines make up the stick figure diagrams. Corresponding to the line configurations, the degree of belongingness of each line is calculated.

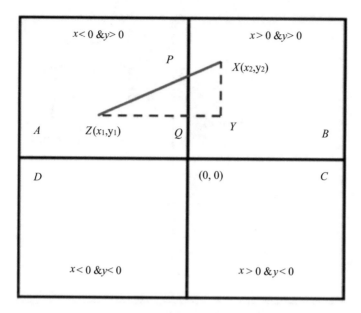

**Fig. 3.9** Similarity triangle creation for fuzzy membership value calculation

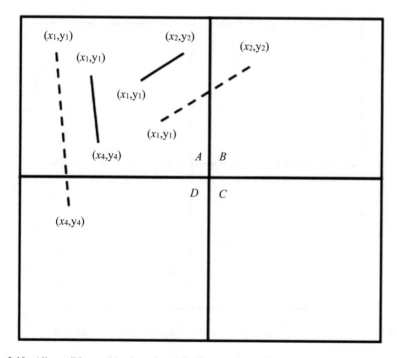

**Fig. 3.10** All possible combination of straight lines corresponding to quadrant 1

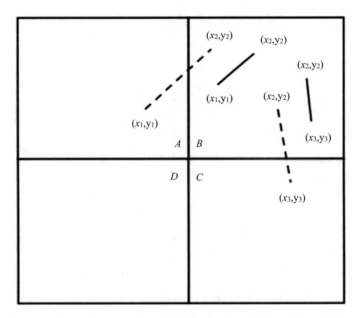

**Fig. 3.11** All possible combination of straight lines corresponding to quadrant 2

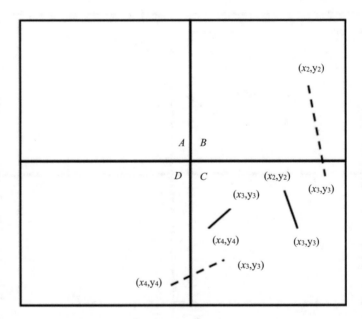

**Fig. 3.12** All possible combination of straight lines corresponding to quadrant 3

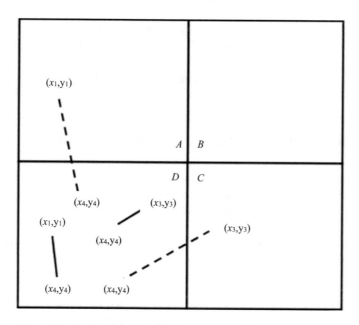

**Fig. 3.13** All possible combination of straight lines corresponding to quadrant 4

For each straight line,

$$\sum_{x=A,B,C,D} \mu_x^{line} = 1 \tag{3.40}$$

$$\mu_A^{line} = \begin{cases} 1 & \text{if } x_1, x_2, x_4 < 0 \,\&\, y_1, y_2, y_4 > 0 \\ \frac{x_1}{x_1 - x_2} & \text{if } x_2, y_1, y_2 > 0 \,\&\, x_1 < 0 \\ \frac{y_1}{y_1 - y_4} & \text{if } x_1, x_4, y_4 < 0 \,\&\, y_1 > 0 \\ 0 & \text{otherwise} \end{cases} \tag{3.41}$$

$$\mu_B^{line} = \begin{cases} 1 & \text{if } x_1, x_2, x_3, y_1, y_2, y_3 > 0 \\ \frac{x_2}{x_2 - x_1} & \text{if } x_2, y_1, y_2 > 0 \,\&\, x_1 < 0 \\ \frac{y_2}{y_2 - y_3} & \text{if } x_2, x_3, y_2 > 0 \,\&\, y_3 < 0 \\ 0 & \text{otherwise} \end{cases} \tag{3.42}$$

$$\mu_C^{line} = \begin{cases} 1 & \text{if } x_2, x_3, x_4 > 0 \,\&\, y_2, y_3, y_4 < 0 \\ \frac{x_3}{x_3 - x_4} & \text{if } x_4, y_3, y_4 < 0 \,\&\, x_3 > 0 \\ \frac{y_3}{y_3 - y_2} & \text{if } x_2, x_3, y_2 > 0 \,\&\, y_3 < 0 \\ 0 & \text{otherwise} \end{cases} \tag{3.43}$$

$$\mu_D^{line} = \begin{cases} 1 & \text{if } x_1, x_3, x_4, y_1, y_3, y_4 < 0 \\ \frac{x_4}{x_4 - x_3} & \text{if } x_4, y_3, y_4 < 0 \ \& \ x_3 > 0 \\ \frac{y_4}{y_4 - y_1} & \text{if } x_1, x_4, y_4 < 0 \ \& \ y_1 > 0 \\ 0 & \text{otherwise} \end{cases} \tag{3.44}$$

## 3.13  Matching of Postures Using Fuzzy T-Norms

Depending on the constituents of the stick figure diagrams, the primitives in two major categories are differentiated, one consisting of only unbounded regions and the other consisting both unbounded and bounded contours. While considering the stick figure diagrams consisting of only unbounded regions the authors find that the posture "posture side" and "attitude back" consists of only 3 and 4 straight lines respectively. These two postures belong to a subcategory within the primitives containing only unbounded regions. Figure 3.14 provides a flowchart demonstrating the generated categories and subcategories. Unknown postures are compared with the primitives which belong to its category. Partitioning the primitives into categories and sub-categories introduces a hierarchical search strategy leading to lowered computational complexity.

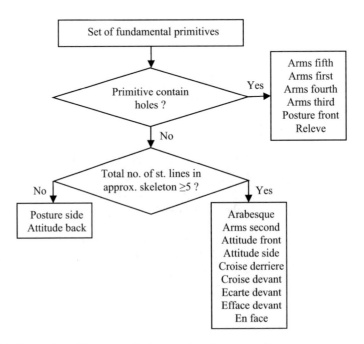

**Fig. 3.14** Partitioning of fundamental primitives into three categories

As discussed earlier in the fuzzy set operations section triangular norms and conforms (S-norms) are used to define a generalized intersection and union of fuzzy sets. Triangular norms and co-norms serve as aggregation operators, which is used to compute the resulting degree of confidence in a hypothesis. The authors subject each of the known postures and the posture to be determined is subjected to the steps discussed in Sects. 3.7–3.12. On determination of the significant straight lines and their corresponding fuzzy membership values of the postures, the authors subject them to Eq. 3.45. Equation 3.46 describes the T-norm operator, with the help of which the authors determine the matching proximity of an unknown Ballet posture to one of the seventeen fundamental dance primitives.

The significant straight lines as mentioned earlier are in a sorted order. Corresponding to each significant straight line the authors obtain a quadtuple of membership values (4 membership values), that is its degree of belongingness to the 4 quadrants. While evaluating the proximity of an unknown posture to a known one, the authors compare their quadtuples. Quadtuples in an unknown posture, that are similar to quadtuples in a known posture, and their relative position less than equal to three, cancel each other, the remaining $n$ straight lines are considered. While the term $q$ signifies the 4 quadrants, $\mu_{ql}^{line\_unknown}$ is the fuzzy membership value of the $l$th line in the $q$th quadrant for an unknown posture, while $\mu_{ql}^{line\_known}$ suggests the same for a known primitive. The calculation of $\mu_{x}^{line}$ is illustrated in Sect. 3.12 $\underset{dp}{T}\left(\mu_{ql}^{line\_unknown}, \mu_{ql}^{line\_known}\right)$ determines the drastic product of the membership values under consideration, thereby calculating the $T$ norm value. The equation governing $\underset{dp}{T}\left(\mu_{ql}^{line\_unknown}, \mu_{ql}^{line\_known}\right)$ is given in (3.46). In (3.45), $M$ calculates the summation of drastic products between the corresponding fuzzy membership values of an unknown posture and the known primitives that belong to its category and finally returns the index number of the primitive that best resembles the unknown posture.

$$M = \underset{\substack{\forall \\ primitives}}{\arg\min} \left[ \sum_{q=A,B,C,D}^{n} \left\{ \sum_{l=1} \underset{dp}{T}\left(\mu_{ql}^{line\_unknown}, \mu_{ql}^{line\_known}\right) \right\} \right] \qquad (3.45)$$

where

$$\underset{dp}{T}\left(\mu_{ql}^{line\_unknown}, \mu_{ql}^{line\_cknown}\right) = \begin{cases} \mu_{ql}^{line\_unknown} & \text{if} & \mu_{ql}^{line\_known} = 1 \\ \mu_{ql}^{line\_known} & \text{if} & \mu_{ql}^{line\_unknown} = 1 \\ 0 & \text{if} & \mu_{ql}^{line\_unknown}, \mu_{ql}^{line\_unknown} < 1 \end{cases}$$

$$(3.46)$$

## 3.14  Algorithm

**Step 0**  Create an initial database with significant lines, stored in sorted order and their corresponding $\mu_x^{line\_known}$ (where $x = A, B, C, D$) of the 17 basic dance primitives of Ballet. Categorize the primitives as postures containing bounded and unbounded regions

*BEGIN*

**Step 1**  Determine the significant straight lines (6 for postures containing bounded regions and 5 for postures containing unbounded regions, with an exception of 3 significant straight lines for postures constituting of less than 5 straight lines) and their corresponding $\mu_x^{line\_unknown}$ (where $x = A, B, C, D$) for the unknown posture whose proximity to the known postures the authors wish to determine.

**Step 2**  *If* the unknown posture contain bounded regions

**2A.**  *Compare the unknown posture with dance primitives containing bounded regions using the following sub-steps:*

*For* known postures $p = 1$–6 containing bounded regions do

*For* significant straight lines $n' = 1$–6 do

Compare the quadtuples, $(\mu_A^{line\_unknown}, \mu_B^{line\_unknown}, \mu_C^{line\_unknown}, \mu_D^{line\_unknown})$ of line $n$ with the $n', n' + 1, n' + 2$ quadtuples $(\mu_A^{line\_known}, \mu_B^{line\_known}, \mu_C^{line\_known}, \mu_D^{line\_known})$ of known posture $p$. Once a match is found, remove the quadtuples from both the known and unknown postures.

*End For*

Compute the number of unmatched quadtuples and store it in $n$
*For* remaining significant straight lines $l = 1$ to $n$ do
*For* quadrants $q = A, B, C, D$ do

a.  Calculate $k = T_{dp}(\mu_{ql}^{line\_unknown}, \mu_{ql}^{line\_known})$ where $T_{dp}$, the fuzzy T-norm operator computes the drastic product between its constituents.

b.  *Sum = Sum + k*

*End For*

*End For*

*End For*

   $M$ holds the index number of the posture with least *Sum* value, and hence the known posture that best matches the unknown posture.

*else*

    `If`

        `The unknown posture contain unbounded regions with   number of sig-`
`nificant lines < 5`

***2B1.*** *Compare the unknown posture with dance primitives containing unbounded regions adhering to the same subsets mentioned in 2A, except by altering the:*

    `(i) number of postures p = 1 to 2 and`
    `(ii)number of significant lines n' = 1 to 3.`
    `else`

***2B2.*** *Compare the unknown posture with dance primitives containing unbounded regions adhering to the same sub steps mentioned in 2A, except by altering the:*

    `(i) number of postures p = 1 to 9 and`
    `(ii) number of significant lines n' = 1 to 5.`
    `End If`

*End If*
*END*

## 3.15   Database Creation

A database is populated with information corresponding to the seventeen basic dance primitives of Ballet as shown in Table 3.1. This database consist of the name of the posture, the original RGB image, the significant lines with respect to the stick figure diagrams and their fuzzy membership values in the four different quadrants. When an unknown posture is chosen for evaluation, its fuzzy membership values are compared with the values of the fuzzy membership values of the seventeen postures present in this database.

## 3.16   Experimental Results

The desideratum of the proposed algorithm is to identify an unknown posture with the 17 fundamental primitives of Ballet. Tables 3.2 and 3.3 illustrates the output corresponding to a known and unknown posture adhering to the algorithm at

**Table 3.1** Database comprising the 17 primitive Ballet postures

| Name | Original image | Significant lines | Fuzzy membership values | | | |
|---|---|---|---|---|---|---|
| Arabesque | | | 0 | 0 | 1.0000 | 0 |
| | | | 0 | 0 | 0 | 1.0000 |
| | | | 1.0000 | 0 | 0 | 0 |
| | | | 0 | 0 | 0 | 1.0000 |
| | | | 0 | 0 | 1.0000 | 0 |
| Arms Fifth | | | 0 | 0 | 1.0000 | 0 |
| | | | 0 | 1.0000 | 0 | 0 |
| | | | 0 | 0 | 0 | 1.0000 |
| | | | 0.3636 | 0.6364 | 0 | 0 |
| | | | 0 | 1.0000 | 0 | 0 |
| | | | 0 | 0 | 0.2083 | 0.7917 |
| Arms First | | | 1.0000 | 0 | 0 | 0 |
| | | | 0.0526 | 0.5 | 0.4474 | 0 |
| | | | 0 | 0 | 0 | 1.0000 |
| | | | 0 | 0 | 1.0000 | 0 |
| | | | 0.5833 | 0 | 0 | 0.4167 |
| | | | 0 | 0 | 0 | 1.0000 |
| Arms Fourth | | | 0 | 0 | 0.9355 | 0.0645 |
| | | | 0 | 1.0000 | 0 | 0 |
| | | | 0.7778 | 0.0000 | 0 | 0 |
| | | | 0 | 1.0000 | 0 | 0 |
| | | | 0 | 0 | 0 | 1.0000 |
| | | | 0 | 0 | 0 | 1.0000 |
| Arms Second | | | 0 | 0 | 0.4020 | 0.5980 |
| | | | 0 | 0 | 1.0000 | 0 |
| | | | 0 | 1.0000 | 0 | 0 |
| | | | 0 | 1.0000 | 0 | 0 |
| | | | 0 | 1.0000 | 0 | 0 |
| | | | 0 | 1.0000 | 0 | 0 |

(continued)

**Table 3.1**  (continued)

| Name | Original image | Significant lines | Fuzzy membership values | | | |
|------|----------------|-------------------|---|---|---|---|
| Arms Third | | | 0 | 0 | 0.4020 | 05980 |
| | | | 0 | 0 | 1.0000 | 0 |
| | | | 0 | 1.0000 | 0 | 0 |
| | | | 0 | 1.0000 | 0 | 0 |
| | | | 0 | 1.0000 | 0 | 0 |
| | | | 0 | 1.0000 | 0 | 0 |
| Attitude Back | | | 0 | 1.0000 | 0 | 0 |
| | | | 0 | 0 | 1.0000 | 0 |
| | | | 0.8571 | 0.1429 | 0 | 0 |
| | | | 0 | 0 | 1.0000 | 0 |
| Attitude Front | | | 1.0000 | 0 | 0 | 0 |
| | | | 0.5311 | 0.2197 | | |
| | | | 0.2492 | 0 | | |
| | | | 0 | 0.5556 | 0.4444 | 0 |
| | | | 0 | 0 | 0 | 1.0000 |
| | | | 0 | 0 | 0 | 1.0000 |
| Attitude Side | | | 0 | 0 | 0.1310 | 0.8690 |
| | | | 0 | 1.0000 | 0 | 0 |
| | | | 0.0294 | 0 | 0 | 0.9706 |
| | | | 0 | 0 | 1.0000 | 0 |
| | | | 0 | 1.0000 | 0 | 0 |
| Croise Derriere | | | 0 | 0 | 0.5200 | 0.4800 |
| | | | 0 | 0 | 1.0000 | 0 |
| | | | 0 | 1.0000 | 0 | 00 |
| | | | 0 | 1.0000 | 0 | 0 |
| | | | 0 | 0 | 1.0000 | 0 |

(continued)

**Table 3.1** (continued)

| Name | Original image | Significant lines | Fuzzy membership values | | | |
|---|---|---|---|---|---|---|
| Croise Devant | | | 0 | 0 | 0.3929 | 0.6071 |
| | | | 0 | 0 | 1.0000 | 0 |
| | | | 0 | 1.0000 | 0 | 0 |
| | | | 0 | 1.0000 | 0 | 0 |
| | | | 0 | 1.0000 | 0 | 0 |
| Ecarte Devant | | | 1.0000 | 0 | 0 | 0 |
| | | | 1.0000 | 0 | 0 | 0 |
| | | | 0 | 0 | 0.8806 | 0.1194 |
| | | | 0 | 0 | 0 | 1.0000 |
| | | | 0 | 0 | 0 | 1.0000 |
| Efface Devant | | | 0.3714 | 0.6286 | 0 | 0 |
| | | | 0 | 1.0000 | 0 | 0 |
| | | | 0 | 0 | 0.4340 | 0.5660 |
| | | | 0 | 0 | 1.0000 | 0 |
| | | | 0 | 0 | 1.0000 | 0 |
| En Face | | | 0 | 0 | 0.0395 | 0.9605 |
| | | | 0 | 0 | 1.0000 | 0 |
| | | | 1.0000 | 0 | 0 | 0 |
| | | | 1.0000 | 0 | 0 | 0 |
| | | | 1.0000 | 0 | 0 | 0 |
| Posture Front | | | 0 | 0 | 0.9487 | 0.0513 |
| | | | 1.0000 | 0 | 0 | 0 |
| | | | 0 | 1.0000 | 0 | 0 |
| | | | 0 | 0.1042 | 0.8958 | 0 |
| | | | 0 | 0 | 0 | 1.0000 |
| | | | 0.0638 | 0 | 0 | 0.9362 |

(continued)

**Table 3.1**  (continued)

| Name | Original image | Significant lines | Fuzzy membership values | | | |
|---|---|---|---|---|---|---|
| Posture Side | | | 0 | 1.0000 | 0 | 0 |
| | | | 0 | 0 | 1.0000 | 0 |
| | | | 0 | 0.7500 | 0.2500 | 0 |
| Releve | | | 1.0000 | 0 | 0 | 0 |
| | | | 0 | 0 | 0 | 1.0000 |
| | | | 0 | 0 | 1.0000 | 0 |
| | | | 0 | 0 | 1.0000 | 0 |
| | | | 0 | 0 | 0.2000 | 0.8000 |
| | | | 0 | 0 | 0 | 1.0000 |

various stages. The table portrays the results generated for an unknown posture, and simultaneously it also projects the values/images generated for the known posture 'ecarte devant' with which it is correctly matched with, by the proposed procedure. The input to the algorithm is an image of the unknown posture in a RGB format. As an initial step, skin color segmentation is performed on the image. The output of which is dilated in order to eliminate the unnecessary aberrations. The output from this step is used to generate the minimized skeleton of the dance posture. Once skeletonization is performed, spurring is done to remove the unnecessary aberrations from the minimized skeleton. The number of pixels spurred is 16 for all instances. Next, straight line approximation is performed on the minimized skeleton. The straight lines corresponding to the connected components are sorted according to their Euclidean distance, and the first 5 lines are considered as significant straight lines. Their corresponding fuzzy membership values with respect to each quadrant are calculated. Once the authors get the fuzzy membership values of the significant straight lines, they are compared with the existing 17 primitives. Figure 3.15 presents the series of steps followed during the matching procedure. The considered unknown posture didn't contain bounded structures and as the number of significant lines is equal to 5, it was compared with the postures

**Table 3.2** Comparison of an unknown image and known posture: 'Ecarte Devant'

| Known image | | |
|---|---|---|
| RGB Image | Skin color segmentation (after Dilation) | Minimized Skeleton |
| | | |
| Approximated straight lines | | Significant straight lines |
| | | |

containing only unbounded regions and having significant lines equal to 5. The output of (3.45) is the index corresponding to 'ecarte devant' in the initial database. Table 3.4 compares the values for the known and unknown postures.

The overall recognition rate of the proposed algorithm is 82.35% where it recognized 14 out of 17 unknown postures. More specifically 11 out of 13 (84.6%)

**Table 3.3** Comparison of an unknown image and known posture: 'Ecarte Devant'

| Unknown image | | |
| --- | --- | --- |
| RGB Image | Skin color segmentation (after Dilation) | Minimized skeleton |
| | | |
| Approximated straight lines | | Significant straight lines |
| | | |

unknown postures containing unbounded regions and 3 out of 4 (75%) unknown postures containing bounded regions are correctly identified. The average computation time is 3.235 s for each frame in an Intel Pentium Dual-Core Processor with non-optimized Matlab implementation. The codes are written in Matlab R2011b.

**Fig. 3.15** The steps followed during the matching procedure

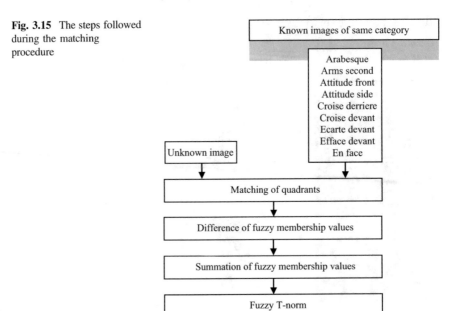

**Table 3.4** Comparison of values between 'Ecarte Devant' posture and unknown posture

| Features | Ecarte Devant | Unknown Posture |
|---|---|---|
| Length of all Straight Lines in descending order | 98.0051, 74.0068, 73.0068, 60.1332, 21.0000, 6.0000 | 133.4541, 130.6484, 112.2185, 105.1190, 30.1496, 26.1725, 25.7099, 14.2127, 12.5300 |
| Fuzzy membership values | $\begin{bmatrix} 1 & 0 & 0 & 0 \\ 1 & 0 & 0 & 0 \\ 0 & 0 & 0.8806 & 0.1194 \\ 0 & 0 & 0 & 1 \\ 0 & 0 & 0 & 1 \end{bmatrix}$ | $\begin{bmatrix} 0 & 0 & 0.8268 & 0.1732 \\ 1 & 0 & 0 & 0 \\ 1 & 0 & 0 & 0 \\ 0 & 0 & 0 & 1 \\ 0 & 0 & 0 & 1 \end{bmatrix}$ |
| Min. sum of Fuzzy T-norm | 0 | |
| Index returned | 12 | |

## 3.17  Conclusion

The procedure stated in this chapter uniquely deals with a wide range of critical dance postures. The entire endeavor proves cost effective as a single static camera can produce the necessary input images for the proposed algorithm. The proposed algorithm is independent of the body type, height and weight of the Ballet dancer, and hence provides even more flexibility to the e-learning process. The algorithm simultaneously addresses the problem of posture recognition and determination of correctness of a particular posture, thereby enhancing the effectiveness of the proposed procedure. The proposed algorithm deals with intricate dance postures, specifically with 17 fundamental dance primitives and registers an accuracy of 84.6% for postures containing unbounded regions, and performs with an overall recognition rate of 82.35%. Considering the complexity of the dance postures, an average computation time of 3.235 s in an Intel Pentium Dual Core processor running Matlab R2011b is highly effective.

To conclude, posture recognition of Ballet dance based on fuzzy posture matching may be considered as a relatively unexplored application area, and the proposed work is an attempt to address the problem with commendable accuracy and scopes for further research.

The proposed procedure is not rotation invariant, therefore the performer must perform in a plane which is parallel to the axis of the camera. The input images to the algorithm need to be perfectly centered, non-adherence to which could produce erroneous results. A closer look at the experimental results would reveal that the performance of the algorithm drops in case of postures containing bounded regions as some of the postures are almost identical, and the proposed algorithm in some cases fail to differentiate between them. These inadequacies provide us with a lot of scope for further improvement over the proposed algorithm.

## 3.18  Matlab Codes

```
% Code is written by Anupam Banerjee
% Under the guidance of Prof. Amit Konar
%this code is applicable for posture which contain holes
%'final matrix' contains 8 columns
%First four columns contains x1, y1, x2, y2 co-ordinate values
%5th column for Euclidean length
%6th column for angle
%7th column for normalized length
%8-11th column shows the fuzzy membership value
temp1=0;
temp2=0;
flag=0;
poly=imread('postureFront_m.jpg');
a=im2bw(poly);
a=~a;
b=bwmorph(a,'spur',14);
c=b;
brow=size(b,1);
```

```
bcol=size(b,2);
row=size(a,1);
column=size(a,2);
for i=1:row
    for j=1:column
        if(b(i,j)==1)
            startrow=i;
            startcolumn=j;
            flag=1;
            break;
        end
    end
    if(flag==1)
        break;
    end
end
%constructing the head part
currx=startrow;
curry=startcolumn;
nextx=0;
nexty=0;
prevx=0;
prevy=0;
temp=0;
k=0;
while(temp==0)
    if (b(currx-1,curry-1)==1)&&((currx-1)~=prevx||(curry-1)~=prevy)
        nextx=currx-1;
        nexty=curry-1;
        k=k+1;
    end
    if (b(currx-1,curry)==1)&&((currx-1)~=prevx||(curry)~=prevy)
        nextx=currx-1;
        nexty=curry;
        k=k+1;
    end
    if (b(currx-1,curry+1)==1)&&((currx-1)~=prevx||(curry+1)~=prevy)
        nextx=currx-1;
        nexty=curry+1;
        k=k+1;
    end
    if (b(currx,curry+1)==1)&&((currx)~=prevx||(curry+1)~=prevy)
        nextx=currx;
        nexty=curry+1;
        k=k+1;
```

```
        end
        if (b(currx+1,curry)==1)&&((currx+1)~=prevx||(curry)~=prevy)
          nextx=currx+1;
          nexty=curry;
          k=k+1;
        end
        if (b(currx+1,curry+1)==1)&&((currx+1)~=prevx||(curry+1)~=prevy)
          nextx=currx+1;
          nexty=curry+1;
          k=k+1;
        end
        if (b(currx,curry-1)==1)&&((currx)~=prevx||(curry-1)~=prevy)
          nextx=currx;
          nexty=curry-1;
          k=k+1;
        end
        if (b(currx+1,curry-1)==1)&&((currx+1)~=prevx||(curry-1)~=prevy)
          nextx=currx+1;
          nexty=curry-1;
          k=k+1;
        end
        if(k==1)
          prevx=currx;
          prevy=curry;
          currx=nextx;
          curry=nexty;
        end
        if(k==2)
          b(prevx,prevy)=0;
          brpoint1x=currx;
          brpoint1y=curry;
          currx=nextx;
          curry=nexty;
          break;
        end
        k=0;
    end
end
plotin=1;
plotindex(plotin,1)=startrow;
plotindex(plotin,2)=startcolumn;
plotin=plotin+1;
plotindex(plotin,1)=brpoint1x;
plotindex(plotin,2)=brpoint1y;
plotin=plotin+1;
plotindex(plotin,1)=0;
```

```
plotindex(plotin,2)=0;
plotin=plotin+1;
%construction of 1st part of the polygon
counter=1;
nextx=0;
nexty=0;
prevx=brpoint1x;
prevy=brpoint1y;
k=0;
while(temp==0)
    if (b(currx+1,curry)==1)&&((currx+1)~=prevx||(curry)~=prevy)
      nextx=currx+1;
      nexty=curry;
      k=k+1;
    end
    if (b(currx-1,curry-1)==1)&&((currx-1)~=prevx||(curry-1)~=prevy)
      nextx=currx-1;
      nexty=curry-1;
      k=k+1;
    end
    if (b(currx-1,curry)==1)&&((currx-1)~=prevx||(curry)~=prevy)
      nextx=currx-1;
      nexty=curry;
      k=k+1;
    end
    if (b(currx-1,curry+1)==1)&&((currx-1)~=prevx||(curry+1)~=prevy)
      nextx=currx-1;
      nexty=curry+1;
      k=k+1;
    end
    if (b(currx,curry+1)==1)&&((currx)~=prevx||(curry+1)~=prevy)
      nextx=currx;
      nexty=curry+1;
      k=k+1;
    end
    if (b(currx+1,curry+1)==1)&&((currx+1)~=prevx||(curry+1)~=prevy)
      nextx=currx+1;
      nexty=curry+1;
      k=k+1;
    end
    if (b(currx,curry-1)==1)&&((currx)~=prevx||(curry-1)~=prevy)
      nextx=currx;
      nexty=curry-1;
      k=k+1;
    end
```

```
  if (b(currx+1,curry-1)==1)&&((currx+1)~=prevx||(curry-1)~=prevy)
    nextx=currx+1;
    nexty=curry-1;
    k=k+1;
  end
  if(k==1)
    polyindex(counter,1)=currx;
    polyindex(counter,2)=curry;
    polyindex(counter,3)=counter;
    counter=counter+1;
    prevx=currx;
    prevy=curry;
    currx=nextx;
    curry=nexty;
  end
  if(k==2)
    brpoint2x=currx;
    brpoint2y=curry;
    prevx=currx;
    prevy=curry;
    currx=nextx;
    curry=nexty;
    counter=counter+1;
    break;
  end
  k=0;
end
p1=round(counter/3);
p2=round(counter/1.5);
plotindex(plotin,1)=brpoint1x;
plotindex(plotin,2)=brpoint1y;
plotin=plotin+1;
for i=1:(counter-2)
  if(polyindex(i,3)==p1)
    plotindex(plotin,1)=polyindex(i,1);
    plotindex(plotin,2)=polyindex(i,2);
    temp1=polyindex(i,1);
    temp2=polyindex(i,2);
    plotin=plotin+1;
    plotindex(plotin,1)=0;
    plotindex(plotin,2)=0;
    plotin=plotin+1;
  end
  if(polyindex(i,3)==p2)
    plotindex(plotin,1)=temp1;
```

```
        plotindex(plotin,2)=temp2;
        plotin=plotin+1;
        plotindex(plotin,1)=polyindex(i,1);
        plotindex(plotin,2)=polyindex(i,2);
        temp1=polyindex(i,1);
        temp2=polyindex(i,2);
        plotin=plotin+1;
        plotindex(plotin,1)=0;
        plotindex(plotin,2)=0;
        plotin=plotin+1;
        break;
    end
end
plotindex(plotin,1)=temp1;
plotindex(plotin,2)=temp2;
plotin=plotin+1;
plotindex(plotin,1)=brpoint2x;
plotindex(plotin,2)=brpoint2y;
plotin=plotin+1;
plotindex(plotin,1)=0;
plotindex(plotin,2)=0;
plotin=plotin+1;
%construction of 2nd half of the hexagon.
counter=1;
nextx=0;
nexty=0;
k=0;
while(temp==0)
    if (b(currx+1,curry)==1)&&((currx+1)~=prevx||(curry)~=prevy)
      nextx=currx+1;
      nexty=curry;
      k=k+1;
    end
    if (b(currx-1,curry-1)==1)&&((currx-1)~=prevx||(curry-1)~=prevy)
      nextx=currx-1;
      nexty=curry-1;
      k=k+1;
    end
    if (b(currx-1,curry)==1)&&((currx-1)~=prevx||(curry)~=prevy)
      nextx=currx-1;
      nexty=curry;
      k=k+1;
    end
    if (b(currx-1,curry+1)==1)&&((currx-1)=prevx||(curry+1)~=prevy)
      nextx=currx-1;
```

```
      nexty=curry+1;
      k=k+1;
    end
    if (b(currx,curry+1)==1)&&((currx)~=prevx||(curry+1)~=prevy)
      nextx=currx;
      nexty=curry+1;
      k=k+1;
    end
    if (b(currx+1,curry+1)==1)&&((currx+1)~=prevx||(curry+1)~=prevy)
      nextx=currx+1;
      nexty=curry+1;
      k=k+1;
    end
    if (b(currx,curry-1)==1)&&((currx)~=prevx||(curry-1)~=prevy)
      nextx=currx;
      nexty=curry-1;
      k=k+1;
    end
    if (b(currx+1,curry-1)==1)&&((currx+1)~=prevx||(curry-1)~=prevy)
      nextx=currx+1;
      nexty=curry-1;
      k=k+1;
    end
    if(k==1)
      polyindex(counter,1)=currx;
      polyindex(counter,2)=curry;
      polyindex(counter,3)=counter;
      counter=counter+1;
      prevx=currx;
      prevy=curry;
      currx=nextx;
      curry=nexty;
    end
    if((currx==brpoint1x)&&(curry==brpoint1y))
      break;
    end
    k=0;
end
p1=round(counter/3);
p2=round(counter/1.5);
plotindex(plotin,1)=brpoint2x;
plotindex(plotin,2)=brpoint2y;
plotin=plotin+1;
for i=1:(counter-2)
    if(polyindex(i,3)==p1)
```

```
            plotindex(plotin,1)=polyindex(i,1);
            plotindex(plotin,2)=polyindex(i,2);
            temp1=polyindex(i,1);
            temp2=polyindex(i,2);
            plotin=plotin+1;
            plotindex(plotin,1)=0;
            plotindex(plotin,2)=0;
            plotin=plotin+1;
        end
        if(polyindex(i,3)==p2)
            plotindex(plotin,1)=temp1;
            plotindex(plotin,2)=temp2;
            plotin=plotin+1;
            plotindex(plotin,1)=polyindex(i,1);
            plotindex(plotin,2)=polyindex(i,2);
            temp1=polyindex(i,1);
            temp2=polyindex(i,2);
            plotin=plotin+1;
            plotindex(plotin,1)=0;
            plotindex(plotin,2)=0;
            plotin=plotin+1;
            break;
        end
    end
plotindex(plotin,1)=temp1;
plotindex(plotin,2)=temp2;
plotin=plotin+1;
plotindex(plotin,1)=brpoint1x;
plotindex(plotin,2)=brpoint1y;
plotin=plotin+1;
plotindex(plotin,1)=0;
plotindex(plotin,2)=0;
plotin=plotin+1;
 %construction of the bottom part!
prevx=(brpoint2x+1);
prevy=brpoint2y;
currx=(brpoint2x+2);
curry=brpoint2y;
nextx=0;
nexty=0;
temp=0;
k=0;
while(temp==0)
    if (b(currx-1,curry-1)==1)&&((currx-1)~=prevx||(curry-1)~=prevy)
        nextx=currx-1;
```

```matlab
      nexty=curry-1;
      k=k+1;
   end
   if (b(currx-1,curry)==1)&&((currx-1)~=prevx||(curry)~=prevy)
      nextx=currx-1;
      nexty=curry;
      k=k+1;
   end
   if (b(currx-1,curry+1)==1)&&((currx-1)~=prevx||(curry+1)~=prevy)
      nextx=currx-1;
      nexty=curry+1;
      k=k+1;
   end
   if (b(currx,curry+1)==1)&&((currx)~=prevx||(curry+1)~=prevy)
      nextx=currx;
      nexty=curry+1;
      k=k+1;
   end
   if (b(currx+1,curry)==1)&&((currx+1)~=prevx||(curry)~=prevy)
      nextx=currx+1;
      nexty=curry;
      k=k+1;
   end
   if (b(currx+1,curry+1)==1)&&((currx+1)~=prevx||(curry+1)~=prevy)
      nextx=currx+1;
      nexty=curry+1;
      k=k+1;
   end
   if (b(currx,curry-1)==1)&&((currx)~=prevx||(curry-1)~=prevy)
      nextx=currx;
      nexty=curry-1;
      k=k+1;
   end
   if (b(currx+1,curry-1)==1)&&((currx+1)~=prevx||(curry-1)~=prevy)
      nextx=currx+1;
      nexty=curry-1;
      k=k+1;
   end
   if(k==1)
      prevx=currx;
      prevy=curry;
      currx=nextx;
      curry=nexty;
   end
   if(k==2)
```

```
        b(prevx,prevy)=0;
        brpoint3x=currx;
        brpoint3y=curry;
        currx=nextx;
        curry=nexty;
        break;
    end
    k=0;
end
plotindex(plotin,1)=brpoint2x;
plotindex(plotin,2)=brpoint2y;
plotin=plotin+1;
plotindex(plotin,1)=brpoint3x;
plotindex(plotin,2)=brpoint3y;
plotin=plotin+1;
plotindex(plotin,1)=0;
plotindex(plotin,2)=0;
plotin=plotin+1;
%constructing the last part!
nextx=0;
nexty=0;
prevx=brpoint3x;
prevy=brpoint3y;
k=0;
counter=1;
flagzero=0;
tempx=currx;
tempy=curry;
while(temp==0)
    if (b(currx+1,curry)==1)&&((currx+1)~=prevx||(curry)~=prevy)
        nextx=currx+1;
        nexty=curry;
        k=k+1;
    end
    if (b(currx-1,curry-1)==1)&&((currx-1)~=prevx||(curry-1)~=prevy)
        nextx=currx-1;
        nexty=curry-1;
        k=k+1;
    end
    if (b(currx-1,curry)==1)&&((currx-1)~=prevx||(curry)~=prevy)
        nextx=currx-1;
        nexty=curry;
        k=k+1;
    end
    if (b(currx-1,curry+1)==1)&&((currx-1)~=prevx||(curry+1)~=prevy)
```

```
      nextx=currx-1;
      nexty=curry+1;
      k=k+1;
   end
   if (b(currx,curry+1)==1)&&((currx)~=prevx||(curry+1)~=prevy)
      nextx=currx;
      nexty=curry+1;
      k=k+1;
   end
   if (b(currx+1,curry+1)==1)&&((currx+1)~=prevx||(curry+1)~=prevy)
      nextx=currx+1;
      nexty=curry+1;
      k=k+1;
   end
   if (b(currx,curry-1)==1)&&((currx)~=prevx||(curry-1)~=prevy)
      nextx=currx;
      nexty=curry-1;
      k=k+1;
   end
   if (b(currx+1,curry-1)==1)&&((currx+1)~=prevx||(curry-1)~=prevy)
      nextx=currx+1;
      nexty=curry-1;
      k=k+1;
   end
   if(k==1)
      polyindex(counter,1)=currx;
      polyindex(counter,2)=curry;
      polyindex(counter,3)=counter;
      counter=counter+1;
      prevx=currx;
      prevy=curry;
      currx=nextx;
      curry=nexty;
   end
   if((currx==brpoint3x)&&(curry==brpoint3y))
      counter=counter+1;
      break;
   end
   if(k==0)
      flagzero=1;
      plotindex(plotin,1)=brpoint3x;
      plotindex(plotin,2)=brpoint3y;
      plotin=plotin+1;
      plotindex(plotin,1)=currx;
      plotindex(plotin,2)=curry;
```

```
        plotin=plotin+1;
        plotindex(plotin,1)=0;
        plotindex(plotin,2)=0;
        plotin=plotin+1;
        break;
    end
    k=0;
end
if(flagzero==0)
    p1=round(counter*(7/20));
    p2=round(counter*(13/20));
    plotindex(plotin,1)=brpoint3x;
    plotindex(plotin,2)=brpoint3y;
    plotin=plotin+1;
    for i=1:(counter-2)
        if(polyindex(i,3)==p1)
            plotindex(plotin,1)=polyindex(i,1);
            plotindex(plotin,2)=polyindex(i,2);
            temp1=polyindex(i,1);
            temp2=polyindex(i,2);
            plotin=plotin+1;
            plotindex(plotin,1)=0;
            plotindex(plotin,2)=0;
            plotin=plotin+1;
        end
        if(polyindex(i,3)==p2)
            plotindex(plotin,1)=temp1;
            plotindex(plotin,2)=temp2;
            plotin=plotin+1;
            plotindex(plotin,1)=polyindex(i,1);
            plotindex(plotin,2)=polyindex(i,2);
            temp1=polyindex(i,1);
            temp2=polyindex(i,2);
            plotin=plotin+1;
            plotindex(plotin,1)=0;
            plotindex(plotin,2)=0;
            plotin=plotin+1;
            break;
        end
    end
    plotindex(plotin,1)=temp1;
    plotindex(plotin,2)=temp2;
    plotin=plotin+1;
    plotindex(plotin,1)=brpoint3x;
    plotindex(plotin,2)=brpoint3y;
```

```
    plotin=plotin+1;
    plotindex(plotin,1)=0;
    plotindex(plotin,2)=0;
    plotin=plotin+1;
end
if(flagzero==1)
    nextx=0;
    nexty=0;
    prevx=tempx;
    prevy=tempy;
    k=0;
    currx=brpoint3x;
    curry=brpoint3y;
    while(temp==0)
       if (b(currx+1,curry)==1)&&((currx+1)~=prevx||(curry)~=prevy)
          nextx=currx+1;
          nexty=curry;
          k=k+1;
       end
       if (b(currx-1,curry-1)==1)&&((currx-1)~=prevx||(curry-1)
~=prevy)
          nextx=currx-1;
          nexty=curry-1;
          k=k+1;
       end
       if (b(currx-1,curry)==1)&&((currx-1)~=prevx||(curry)~=prevy)
          nextx=currx-1;
          nexty=curry;
          k=k+1;
       end
       if (b(currx-1,curry+1)==1)&&((currx-1)~=prevx||(curry+1)
~=prevy)
          nextx=currx-1;
          nexty=curry+1;
          k=k+1;
       end
       if (b(currx,curry+1)==1)&&((currx)~=prevx||(curry+1)~=prevy)
          nextx=currx;
          nexty=curry+1;
          k=k+1;
       end
       if (b(currx+1,curry+1)==1)&&((currx+1)~=prevx||(curry+1)
~=prevy)
          nextx=currx+1;
          nexty=curry+1;
```

```
          k=k+1;
       end
       if (b(currx,curry-1)==1)&&((currx)~=prevx||(curry-1)~=prevy)
          nextx=currx;
          nexty=curry-1;
          k=k+1;
       end
       if (b(currx+1,curry-1)==1)&&((currx+1)~=prevx||(curry-1)
~=prevy)
          nextx=currx+1;
          nexty=curry-1;
          k=k+1;
       end
       if(k==1)
          polyindex(counter,1)=currx;
          polyindex(counter,2)=curry;
          polyindex(counter,3)=counter;
          counter=counter+1;
          prevx=currx;
          prevy=curry;
          currx=nextx;
          curry=nexty;
       end
       if(k==0)
          plotindex(plotin,1)=brpoint3x;
          plotindex(plotin,2)=brpoint3y;
          plotin=plotin+1;
          plotindex(plotin,1)=currx;
          plotindex(plotin,2)=curry;
          plotin=plotin+1;
          plotindex(plotin,1)=0;
          plotindex(plotin,2)=0;
          plotin=plotin+1;
          break;
       end
     k=0;
     end
end
plotline(plotindex,plotin,brow,bcol);
```

**Results:** Sample run results are given in Fig. 3.6.

```
%this code is applicable for postures which does not contain holes
image=imread('croisedevant_m.jpg');
```

```
stindex=zeros(1,1);
temp1=0;
a=im2bw(image);
a=~a;
b=bwmorph(a,'spur',14);
%imshow(b);
row=size(a,1);
column=size(a,2);
CC=bwconncomp(b);
noconn=CC.NumObjects;
rowin=0;
colin=0;
temp=0;
stin=0;%This is the value to increment the index of the matrix that denotes
the additional straight lines.
startpoint=zeros(noconn,2);
for i=1:noconn
    temp=CC.PixelIdxList{1,i}(1,1);
    [rowin,colin]=ind2sub([row,column],temp);
    startpoint(i,1)=rowin;
    startpoint(i,2)=colin;
end
  %the outermost loop to construct the entire image
for f=1:noconn
    currx=startpoint(f,1);
    curry=startpoint(f,2);
    nextx=0;
    nexty=0;
    prevx=0;
    prevy=0;
    temp=0;
    k=0;
    connx3=0;
    conny3=0;
    connx3n=0;
    conny3n=0;
    rectcount=0;
    %construct to find out the 3connected pixel to draw the head location.
    while(temp==0)
    rectcount=rectcount+1;%for the erroneous behaviour of the bwcon-
ncomp function
        if (b(currx-1,curry-1)==1)&&((currx-1)~=prevx||(curry-1)
~=prevy)
            nextx=currx-1;
            nexty=curry-1;
```

```
        k=k+1;
    end
    if (b(currx-1,curry+1)==1)&&((currx-1)~=prevx||(curry+1)
~=prevy)
        nextx=currx-1;
        nexty=curry+1;
        k=k+1;
    end
    if (b(currx,curry-1)==1)&&((currx)~=prevx||(curry-1)~=prevy)
        nextx=currx;
        nexty=curry-1;
        k=k+1;
    end
    if (b(currx,curry+1)==1)&&((currx)~=prevx||(curry+1)~=prevy)
        nextx=currx;
        nexty=curry+1;
        k=k+1;
    end
    if (b(currx+1,curry-1)==1)&&((currx+1)~=prevx||(curry-1)
~=prevy)
        nextx=currx+1;
        nexty=curry-1;
        k=k+1;
    end
    if (b(currx+1,curry)==1)&&((currx+1)~=prevx||(curry)~=prevy)
        nextx=currx+1;
        nexty=curry;
        k=k+1;
    end
    if (b(currx+1,curry+1)==1)&&((currx+1)~=prevx||(curry+1)
~=prevy)
        nextx=currx+1;
        nexty=curry+1;
        k=k+1;
    end
    %Reshuffling of the if conditions to give priority to the head
        construct.
    if (b(currx-1,curry)==1)&&((currx-1)~=prevx||(curry)~=prevy)
        nextx=currx-1;
        nexty=curry;
        k=k+1;
    end
    if(rectcount==1&&k==2)
rectprevx=currx;
rectprevy=curry;
```

```
rectcurrx=currx+1;
rectcurry=curry;
[counting,recti,rectj]=rectify(rectprevx,rectprevy,rectcurrx,rec-
tcurry,b);
if(counting < 10)
    b(currx+1,curry)=0;
    k=0;
    continue;
end
if(counting >=10)
    currx=recti;
    curry=rectj;
    startpoint(f,1)=recti;
    startpoint(f,2)=rectj;
    k=0;
    continue;
end
        end;
        prevx=currx;
        prevy=curry;
        currx=nextx;
        curry=nexty;
        %Construct to find the coordinates of the end head pixel, and to dis-
connect the pixel that was making the pixel in discussion a 3 con-
nected one.
        if(k==2)
            connx3=prevx;
            conny3=prevy;
            b(currx,curry)=0;%Disconnects the pixel.
            %disp('disconnected');
            %disp(currx);
            %disp(curry);
            temp1=0;
            k1=0;
            while(temp1==0)
                if (b(currx-1,curry-1)==1)&&((currx-1) ~=prevx||(curry-1)
~=prevy)
                    prevx=currx;
                    prevy=curry;
                    currx=currx-1;
                    curry=curry-1;
                    k1=1;
                end
                if (b(currx-1,curry)==1)&&((currx-1) ~=prevx||(curry)
~=prevy)
```

```
        prevx=currx;
        prevy=curry;
        currx=currx-1;
        k1=1;
    end
    if (b(currx-1,curry+1)==1)&&((currx-1)~=prevx||(curry+1)
~=prevy)
        prevx=currx;
        prevy=curry;
        currx=currx-1;
        curry=curry+1;
        k1=1;
    end
    if (b(currx,curry-1)==1)&&((currx)~=prevx||(curry-1)
~=prevy)
        prevx=currx;
        prevy=curry;
        curry=curry-1;
        k1=1;
    end
    if (b(currx,curry+1)==1)&&((currx)~=prevx||(curry+1)
~=prevy)
        prevx=currx;
        prevy=curry;
        curry=curry+1;
        k1=1;
    end
    if (b(currx+1,curry-1)==1)&&((currx+1)~=prevx||(curry-1)
~=prevy)
        prevx=currx;
        prevy=curry;
        currx=currx+1;
        curry=curry-1;
        k1=1;
    end
    if (b(currx+1,curry)==1)&&((currx+1)~=prevx||(curry)
~=prevy)
        prevx=currx;
        prevy=curry;
        currx=currx+1;
        k1=1;
    end
    if (b(currx+1,curry+1)==1)&&((currx+1)~=prevx||(curry+1)
~=prevy)
        prevx=currx;
```

```
                    prevy=curry;
                    currx=currx+1;
                    curry=curry+1;
                    k1=1;
                end
                if(k1==0)
                    connx3n=currx;
                    conny3n=curry;
                    temp1=1;
                    break;
                end
                k1=0;
            end
        end
        if(k==0)
            break;
        end
        if(temp1==1)
            stin=stin+1;
            stindex(stin,1)=connx3;
            stindex(stin,2)=conny3;
            stindex(stin,3)=connx3n;
            stindex(stin,4)=conny3n;
            temp1=0;
            break;
        end
        k=0;
    end
end
straighten1(b,startpoint,stindex,noconn);
```

**Results:** Sample run results are given in Fig. 3.7.

```
loop=5;%no of significant lines
s=size(Ecarte_Devant);
%Ecarte_Devant contains fuzzy membership values for Ecarte Devant posture
t=(s(1))-loop1;
for i=1:t
    s=size(EcarteDevant);
    a=s(1);
    EcarteDevant(a,:)=[];
end
loop1=loop;
unknown1=unknown;
```

```
%unknown contains fuzzy membership values for Unknown posture
flag1=0;
for i=1:loop1
    i;
    for j=loop1:-1:1
        j;
        if EcarteDevant(i,:)==unknown1(j,:)
            EcarteDevant(i,:)=[];
            unknown1(j,:)=[];
            loop1=loop1-1;
            flag1=flag1+1;
        end
    end
    loop1=loop1-1;
end
value=EcarteDevant-unknown1;
value=abs(value);
```

**Results**: Sample run results are given in Table 3.4.

# References

1. D. Travis, *Effective Color Displays: Theory and Practice*, vol. 1991 (Academic Press, London, 1991)
2. P. Maragos, R. Schafer, Morphological skeleton representation and coding of binary images. IEEE Trans. Acoust. Speech Sig. Process. **34**(5), 1228–1244 (1986)
3. J.J. Kotze, J.J.D. Van Schalkwyk, Image coding through skeletonization, in *Proceedings of the 1992 South African Symposium on Communications and Signal Processing, 1992. COMSIG'92* (1992), pp. 87–90
4. D.A. Molloy, *Machine Vision Algorithms in Java: Techniques and Implementation* (Springer, London, 2001)
5. F. Xie, G. Xu, Y. Cheng, Y. Tian, Human body and posture recognition system based on an improved thinning algorithm. Image Process. IET **5**(5), 420–428 (2011)
6. H. Freeman, On the encoding of arbitrary geometric configurations. IRE Trans. Electron. Comput. **2**, 260–268 (1961)
7. A. Konar, *Computational Intelligence: Principles, Techniques and Applications* (Springer, Berlin, 2005)
8. A. Konar, *Artificial Intelligence and Soft Computing: Behavioral and Cognitive Modeling of the Human Brain*, vol. 1 (CRC Press, Boca Raton, 1999)
9. H. Wang, S.-F. Chang, A highly efficient system for automatic face region detection in MPEG video. IEEE Trans. Circuits Syst. Video Technol. **7**(4), 615–628 (1997)

10. B.G. Batchelor, B.E. Marlow, Fast generation of chain code. IEEE Proc. Comput. Digital Tech. **127**(4), 143–147 (1980)
11. E. Bribiesca, A new chain code. Pattern Recogn. **32**(2), 235–251 (1999)
12. H. Sanchez-Cruz, R.M. Rodriguez-Dagnino, Compression bi-level images by means of a 3-bit chain code. SPIE Opt. Eng. **44**, 1–8
13. Y. Kui Liu, B. Žalik, An efficient chain code with Huffman coding. Pattern Recogn. **38**(4), 553–557 (2005)

# Chapter 4
# Gesture Driven Fuzzy Interface System for Car Racing Game

## 4.1 Introduction

The rapid growth of computer applications for normal users in the second half of 20th century called for a robust and reliable system to interact with the machine, thus initiating anew subject area: Human Computer Interaction, also known as HCI. The evolution in input devices to achieve more intricate man-machine interaction with physical feedback has been made possible by the recent advancement in sensors, display and rendering technology. The principle used to recognize the gestures may be based on the mimicry of human vision system: the reception of photons on the retina and the processing of the signal by visual cortex [1]. A line of motion sensing devices has been developed by different corporations that can track the human motion and basic gestures using infrared sensing. The most popular motion sensing devices are Microsoft Kinect, Nintendo Wii, Sony PlayStation Eye and Sega Dreameye. The Microsoft Kinect gives a skeleton of a human standing in its field of view with the coordinate of 20 joints that can be used for gesture recognition algorithms. Equipped with such revolutionary improvement of input devices, computer games, which have been an integral part of computer based entertainment and solely depend of the features offered by HCI devices, have experienced a paradigm shift in developing advanced control by adding more degrees of freedom into the gaming environment. Considering the driving based games, gesture tracking can lead to a new level of experience beyond classical joystick control, enabling the player to move his/her limbs seemingly controlling an imaginary steering wheel or joystick to move through the game. Also the tracking of feet movements would yield speed controls. Keeping in mind the simulation of virtual reality by head-mounted holographic displays of by Oculus Rift or Microsoft Holo Lens, gesture controlled steering seems an indispensible part of driving, eliminating joystick or steering wheel and pedals. However the challenge lies in the fact that gesture-inputs would be ridden with more inaccuracy, redundancy and

© Springer International Publishing AG 2018
A. Konar and S. Saha, *Gesture Recognition*, Studies in Computational
Intelligence 724, DOI 10.1007/978-3-319-62212-5_4

noise than that occur in a classical contact-based input device, and must be processed with intelligent data-processing and recognition algorithms.

The gesture recognition algorithms and applications have been taken the centre field of HCI in the recent decade. The development of an SDK by Microsoft for Kinect sensors has made it an unparalleled motion sensing device used in research and development. Bonanni [2] developed a gesture driven interaction with a mobile robot using Kinect sensors where he employed hidden Markov model [3]. Using of Kinect sensor in hand gesture recognition is carried out by Li [4]. Oszust et al. [5] have proposed a method for recognition of signed expressions observed by the Kinect sensor. Here skeletons of human body with shape and position of hands are taken into account for polish sign language recognition. The recognition is implemented using k-nearest neighbour (kNN) classifier. Burba et al. [6] have used Microsoft Kinect to monitor respiratory rate is estimated by measuring the visual expansion and contraction of the users chest cavity. This work also lightens the area of measuring fidgeting behavior. This is also done by Kinect sensor focusing on vertical oscillations of the user's knees.

The feedback-actuator system for steering wheel driving has been improved and modified with different interfaces in the past years to provide more stability and robustness. Zhang et al. [7] experimented on driving control of electric vehicles and proposed a new H1 robust control strategy. Wada and Kimura [8] analysed the stability of a joystick interface for driving where a steering wheel and gas/brake pedals are actuated by electric motors which rotations are controlled by microcomputers based on 2DOF joystick commands.

Recent works have been focussing on integrating gesture in racing video games, especially in the form of touch-screen gesture or dummy steering wheel. Pfleging et al. [9] has combined both the speech and gesture on a multi-touch steering wheel. Using gestures for manipulation, they provided a fine grained control with immediate feedback. Siversten [10] used a steering wheel input device with pointer event detection and angle compensation technology that can serve as input signals for vehicular control. However, due to the extremely high degree of freedom required by a driving control system, free-air gesture is best suited for it.

The proposed model of driving control for a racing game mimics the natural driving of an auto-transmission car with two hands kept in such a position that they seem holding an imaginary steering wheel. The player sits in a chair and moves his/her right and left feet forward to actuate acceleration and brake. The steering angle is determined by the angle between axis created by joining both hand-coordinates and the transverse plane. As in the free-air gesture, the player may get the perception of linear displacement as the governing quantity and may turn the imaginary wheel in a greater angle when steering radius is small and vice versa, the latter quantity is also taken into account while generating the control signal. The control signal is generated as the keyboard interrupts for "key-down" and "key-up" events with the duration between them as the "key-press duration". The duration is controlled by the steering angle and radius using a type-I Mamdani Fuzzy Inference System to make it precise and robust. The acceleration and brake is controlled by the detection of relative feet position from the skeleton tracked by Kinect.

The main distinction that our driving model presents in the universe of gesture driven racing games is its enhanced reality, more precise and comfortable control of the car movement and the incorporation of acceleration and brake under gesture control. Inspired from the driving system presented in the Forza Motorsport 4 [11] where a Kinect driven mode is also included, we have chosen the normal steering wheel control for our model. However, in the said game the player's gesture is constrained by the fact that he has to keep the hands in right angle with the body-axis for movement, which may be tiresome as referred by the critics [11]. Moreover, the acceleration and brake control is automatic, cutting short the degrees of freedom. We have eliminated that need by considering only the line joining both hand joint coordinates so that the hands can be placed comfortable at any angle with folded elbow-joint. Another major improvement is that in the proposed model acceleration and brake can also be controlled using gesture which was not available for Forza Motorsport 4. The most important of all is that in our model, even the wild-turning of the imaginary steering wheel will not cause disastrous effect as it does in Forza Motorsport 4 since the steering angle is processed by a Fuzzy Inference System (FIS) along with the steering diameter. Another improvement is the additional feature of speed control with feet movements which simulates real-time driving experience.

The rest of the chapter is organised as follows. Section 4.2 gives a brief overview of Kinect for Xbox 360 sensor and the description of its skeleton tracking activity. Section 4.3 gives a vivid analysis of proposed driving model and the insight of the Fuzzy Inference System. We have described the experimental setup in Sect. 4.4. The relevant simulation results and discussion on them constitutes Sect. 4.5. Section 4.6 discusses the future work and concludes the chapter.

## 4.2 Kinect Sensor and Its Specifications

Kinect is a line of motion sensing input devices originally developed for Xbox and Xbox 360 video game consoles and later extended for non-gaming uses in Windows-PC. The Kinect sensor for Xbox 360 [12–14] consists of an RGB camera, an IR camera, an IR projector and a microphone array. It basically looks like a webcam (8-bit VGA resolution of $640 \times 480$ pixels with a Bayer color filter). Its long horizontal bar is provided with a motorized base and tilt. The Kinect sensor along with the associated Software Development Kit (SDK) tracks the human motion by generating a skeleton with three dimensional co-ordinates within a finite range of distance (roughly 1.2–3.5 m or 3.9–11 ft). This is achieved using the visible IR (which is the depth sensor, by which $z$ direction value for each joint is obtained) and RGB cameras. The skeleton produced by the Kinect sensor has twenty body joints. Figure 4.1 shows all the joints along with the index $J_i$ ($1 \leq i \leq 20$) of the human body considered in the skeleton. The background, dress

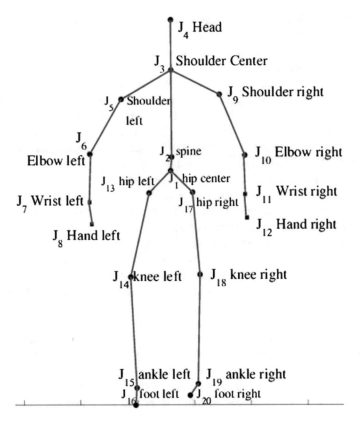

**Fig. 4.1** Twenty body joint obtained using a Kinect sensor

color and lighting of the room are irrelevant for skeleton detection using the Kinect sensor. Hence it can recognize human motion in a very wide range of surrounding physical conditions.

## 4.3   Gesture Controlled Driving Model

The principal objective of gesture controlled driving is to provide a smooth steering control by means of comparatively easy dynamic gesture and control the acceleration and brake of a car by similarly easy gesture potentially unrelated to that used in steering so that they do not interfere. The most intuitive and easiest solution is to mimic the normal gesture used in real life driving. When driving a real automobile vehicle, the direction is controlled by turning a steering wheel with both hands, and the acceleration and brake is actuated by pressing corresponding pedals with right and left feet respectively. The proposed model has adopted this steering wheel concept by tracking the gestures produced by two hands while holding an

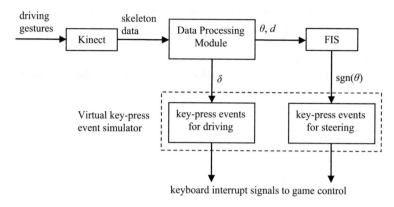

**Fig. 4.2** Block diagram for gesture driven fuzzy interface system

imaginary steering wheel. The angle generated by the straight line connecting two palm position (left and right hand joints in Kinect skeleton) with the transverse or horizontal plane is given as the input to the steering system along with its sign. The overview of the proposed mechanism is shown in the form of block-diagram in Fig. 4.2. The complete algorithm is given in Fig. 4.3.

### 4.3.1   Fuzzy Inference System for Steering

The principal challenge to a gesture driven steering control system is to achieve a desired robustness. The success of such a system depends on the precise detection and measurement of human limb-movements that are noisy and sometimes unintentionally spurious. Additionally there are inherent IF-THEN conditional relationships between input and output variables which are easily identifiable and such existence serves as a motivation to form a fuzzy rule base instead of formulating exact functional mapping. We propose a Type-I Mamdani Fuzzy Inference System (FIS) to generate the steering output for the racing car game. The FIS consists of two input variables and one output variable.

(1) **Input 1: The Angle of Steering ($\theta$):**

The fundamental input for steering a car in the racing game is the steering angle, i.e. the angle by which the player seems to turn the imaginary steering wheel. To get a proper measurement of this angle, we calculated the acute angle between the transverse plane and straight line joining left and right-hand joint-coordinated in the skeleton tracked by the Kinect. Experiment shows that even during the intention extreme turn, this angle doesn't cross the maximum value of 60°, and so the scale of the input steering angle has been calibrated from −60° to +60°. The angle $\theta$ is shown in Fig. 4.4 with the help of a 3D image of a player in driving position.

```
1: procedure EVALFIS(d, θ)
2:        Return the output of the FIS
3: end procedure
4: procedure PRESS KEY(T,keycodek), where k is the virtual key code
   of the corresponding key
5:        key_down(k);
6:        Sleep(T);
7:        key_release(k);
8: end procedure
9: ParBegin
10: procedure STEER
11:        while true do
12:                T ← EVALVIS(d,θ)
13:                ifθ ≥ θ₀then
14:                        ifθ> 0 then
15:                                PRESS_KEY(T, RIGHT_ARROW)
16:                        else
17:                                PRESS_KEY(T, LEFT_ARROW)
18:                        end if
19:                end if
20:        end while
21: end procedure
22: procedure DRIVE
23:        while true do
24:                δ>left_foot.z– right_foot .z
25:                T←k ×δ, wherek is the proportionality constant
26:                ifδ>thresholdthen
27:                        ifδ> 0 then
28:                                PRESS_KEY(T, UP_ARROW)
29:                        else
30:                                PRESS_KEY(T, DOWN_ARROW)
31:                        end if
32:                end if
33:        end while
34: end procedure
35: procedure UPDATE
36:        while true do
37:                draw data from Kinect sensors and processθ, d, and δ
38:        end while
39: end procedure
40: ParEnd
```

**Fig. 4.3** Pseudo-code for the proposed work

**Fig. 4.4**  Three-dimensional
model of a player while
driving

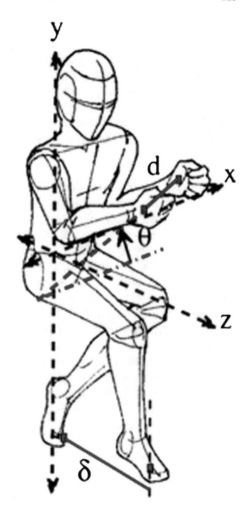

(2) **Input 2: The Steering Diameter** (*d*):

The steering diameter, i.e. the diameter of the steering wheel, is usually fixed during
a real driving system; but in gesture-controlled driving, the analogue of the said
item, the distance between two hand joints, may vary widely. According to the
common behaviour of human sensory-motor system, the driver tends to rotate the
wheel in a greater angle if the diameter is small. But since actual rotation depends
on the steering angle, thus, smaller the diameter of the wheel, smaller is the angular
displacement to produce an equal amount of turn. This poses increasing sensitivity
of rotation as the diameter of wheel decreases. It is to be noted that for small
vehicles like motor-cycle or small cars where frequent large turns are required, the
steering diameter is small, whereas, for larger vehicles and ships, where smaller

(and precise) turn is necessary, a large wheel is provided. But here the intended turning radius of a car must be within a certain range and it must be free from the error coming from human perception of turn due to variable steering diameter. So the steering diameter is also calculated from the Kinect-tracked skeleton and fed to the FIS to calculate the precise Key-press duration. The input to the FIS is a normalized radius for the sake of generalization. The minimum and maximum feasible diameters should be calibrated according to the gamer. The steering diameter $d$ is also shown in Fig. 4.4 with help of a 3D image.

(3) **Steering Output (T)**:

Our principal aim is to generate a signal that controls the direction of the car's motion in gaming environment. In a keyboard controlled racing game, a car rotates as long as the user presses the proper key for e.g., to turn the car at a desired angle at left, the left arrow key is pressed for a proportional duration. This "key-press" event can be virtually simulated by combined "key-down" and "key-up" events which are capable of generating system calls for key-board interrupts. The time duration of "key-press" event which is essentially the time lapse between the instances of "key-down" and "key-up" events related right-arrow or left-arrow key (i.e. the virtual key codes) is the factor to be modulated. The width of the time-pulse, hereby termed as "Key-press duration" $(T)$ is the quantity governing the amount of rotation the car undergoes during the steering movement. In absence of greater level of integration with the game console for the time being, the output of the proposed Fuzzy Inference System (FIS) will provide this quantity to the program producing the virtual key-stroke signals at the user end. We have experimentally determined that the response of the racing game is logarithmic and hence converted the output scale of the FIS into a logarithmic one. It has been determined experimentally that a $T = 200$ ms in a sample is satisfactory as the maximum Key-press duration for sharp bends.

(4) **Fuzzy Rule Base**:

- Inputs: $d, \theta$
- Output: $T$

For $d$, we choose two linguistic constants, viz. Low and High; For $\theta$, Low, Mid and High; and for $T$, we choose Very Low (V. Low), Low, Mid and High. As we need precise control over $T$, it has the maximum number of linguistic constants. The primary input for steering is the steering angle $\theta$ having precedence over steering diameter $d$. So, $\theta$ has more linguistic constants than $d$.

1. IF $d$ is *Low* AND $\theta$ is Low THEN $T$ is *Very Low*.
2. IF $d$ is *Low* AND $\theta$ is *Mid* THEN $T$ is *Low*.
3. IF $d$ is *Low* AND $\theta$ is *High* THEN $T$ is *Mid*.

4. IF $d$ is *High* AND $\theta$ is *Low* THEN $T$ is *Low*.
5. IF $d$ is *High* AND $\theta$ is *Mid* THEN $T$ is *Mid*.
6. IF $d$ is *High* AND $\theta$ is *High* THEN $T$ is *High*.

(5) **Memberships of Input and Output Variables**:

For input variables, Gaussian membership functions have been used. Membership functions of output variable are triangular. Membership function plots for $d$, $\theta$ and $T$ have been shown in Fig. 4.5a–c respectively. The output scale is calibrated in algorithmic way, i.e. the scale is taken as the power of 10. The parameters of membership functions are tabulated in Table 4.1. For Gaussian memberships, mean and standard deviation are given; for triangular ones, three parameters, $\alpha$, $\beta$ and $\gamma$ are given. $\alpha$, $\beta$ and $\gamma$ are the starting threshold, peak and ending threshold of the triangular function. For $\theta$, the scale has been chosen from $\theta_0 = 10°$, instead of $0°$ as the threshold steering angle for actuating key-press is taken at $\pm 10°$. To maintain sharp sigmoid characteristics, the standard deviation for "Low" is chosen much higher than that of "Mid" and "High" for $\theta$.

## 4.3.2   Acceleration and Brake Control

The normal gesture of pressing the pedals with left and right feet is used for acceleration and brake control. The gamer sits in a driving position and puts his/her right foot forward relative to the left foot to signal acceleration of the car. The corresponding "key-press" event is invoked. The gesture is reserved for braking, i.e. left foot has to be advanced with respect to the right foot. Keeping the feet in normal position signals the maintenance of uniform speed. The difference of $z$-coordinates ($\delta$) of right and left feet joints of the skeleton is used to generate the acceleration and deceleration signal. When $\delta$ exceeds a suitable threshold determined experimentally, the car accelerated or decelerates or keeps uniform speed depending on the sign of $\delta$. The signal is actually generated in the form of a keyboard interrupt for up-arrow and down-arrow key and "key-press" duration for them. The key-press duration is proportional to the difference quantity measured from the skeleton. Figure 4.4 gives a clear depiction of $\delta$ in the 3D image of a player.

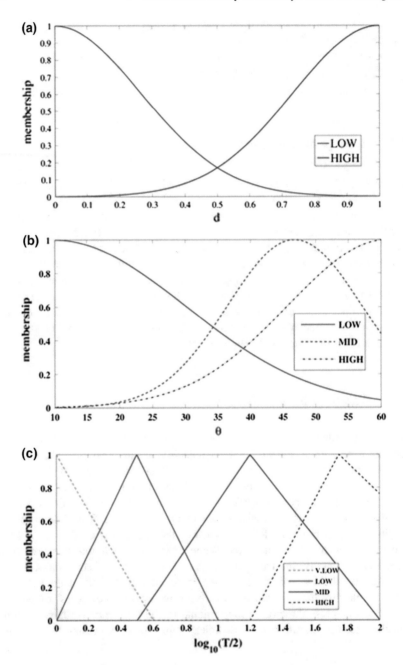

**Fig. 4.5  a** Membership functions obtained for input variable $d$. **b** Membership functions obtained for input variable $\theta$. **c** Membership functions obtained for output variable $T$

**Table 4.1** Experimentally taken membership value parameters

| Variable | Linguistic constant | Mean | Standard deviation | $\alpha$ | $\beta$ | $\gamma$ |
|---|---|---|---|---|---|---|
| $d$ | low | 0 | 0.265 | – | – | – |
| | High | 1 | 0.265 | – | – | – |
| $\theta$ | Low | 9.41 | 25.3 | – | – | – |
| | Mid | 46.77 | 10.3 | – | – | – |
| | High | 61.03 | 15.3 | – | – | – |
| $T$ | Very low | – | – | 0.8 | 0 | 0.6 |
| | Low | – | – | 0 | 0.5 | 1 |
| | Mid | – | – | 0.5 | 1.2 | 2 |
| | High | – | – | 1.2 | 1.75 | 2.88 |

## 4.4 Experimental Setup

To test the efficiency and smoothness of the proposed driving control system, it is applied on a real-time simulator racing game with turns and acceleration function. Final Drive Fury by Wild-Tangent Games has been used as the test bed for the driving-mechanism. This game is ideal to test the mechanism due to its sufficiency of driving inputs in accordance with real-life driving scenario unlike arcade racing games. Moreover the necessary inputs to control this game are brief, explicit and well-defined. Final Drive Fury is a competitive racing game where the player races with others in a cosmopolitan racing track. The control includes acceleration by the up-arrow key, brake and going back by the down-arrow key, and steering using left and right arrows. The game needs Windows XP or above operating system with a minimum requirement of 256 MB RAM. There is no provision of "nitro" or "drift" making the game more realistic. This game needs the rigor and precision of a simulator racing game.

A Windows 7 PC with 2 GB RAM has been used to test the driving control system on this game. A Kinect 360 sensor is used to track the player's skeleton and the project is developed in Visual C++ using Kinect SDK 1.8. The player sits in a comfortable position as shown in the Fig. 4.6 while playing the game. The Kinect is placed at a distance of (1.5–2.1) metre from the player and at a height of (0.4–0.8) metre. The Kinect tracks the skeleton and sends data to the PC where it is processed by FIS and $T$ for steering and acceleration-brake is generated. The player gets the visual information of the game in a screen placed in front of him/her.

The system is tested on twenty subjects as players, all within the age group of 20–30 years. Ten of them were completely inexperienced about any type of simulator racing game before subjected to the test, while six had simulator racing experience beforehand but not with this particular game. Rest were sufficiently habituated with this game. The joints used for this model are enumerated below:

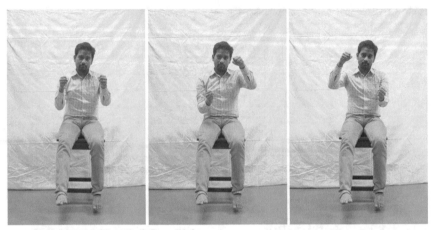

Straight Drive Turn Left Turn Right

**Fig. 4.6** RGB images of a subject while driving the virtual car for gaming purpose

- Left Hand, Right Hand (For measuring steering diameter)
- Shoulder Centre, Spine and Hip Centre (For constructing sagittal plane)
- Left Foot, Right Foot (For acceleration and brake control).

The input steering angle is the angle between the axis joining left and right hand and the transverse plane. But since it's difficult to get transverse plane data from Kinect skeleton, we calculated the angle with the Sagittal plane (plane forming by Shoulder Centre, Spine and Hip Centre) and took its complementary angle as shown in Fig. 4.7.

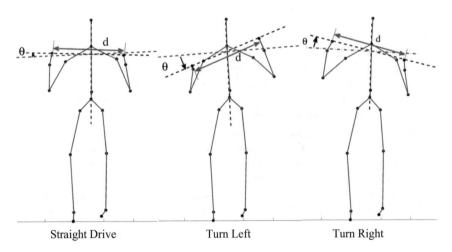

Straight Drive                    Turn Left                    Turn Right

**Fig. 4.7** Tracked skeleton of the player from Fig. 4.6

## 4.5   Results and Discussion

The player sits as shown in Fig. 4.6 while playing the game. The skeleton tracked by the Kinect application is provided in Fig. 4.7 for three principal hand positions while playing. The data generated by Kinect is given for the processing and the processed data in the form of $d$, $\theta$ and $\delta$ are fed to the FIS.

In Table 4.2, the sampled data from Kinect skeleton and the corresponding Key-press duration T and key-press type has been shown. When $\theta < \theta_0$, where $\theta_0$ is set at $10°$, the output $T$ is irrelevant as no keyboard interrupt signal is generated for right or left turn. This ensured no turning with noisy hand movements. The game snapshots for turning left, right and driving straight is shown in Fig. 4.8. It corresponds with the skeleton positions in Fig. 4.7.

The output of the FIS for the range of interest is given in Fig. 4.9 by means of a surface plot. The surface plot shows the sigmoid cross-section for $\theta - \log T$ plane marked by a redline for $d = 0{:}7$. The sigmoid function is necessary for precise control in lower steering angle and sharp bend in higher ones.

To show the non-linearity of the input-output relationship achieved by the FIS, we compare the output with a linear non-Fuzzy system, where $T = c_1 d + c_2 \theta + c_3$. Figure 4.10 illustrates that the plane, corresponding to a particular $(c_1, c_2, c_3)$ which has been obtained by the best linear fit to the surface plot of $T$ obtained from FIS. But as seen from Fig. 4.10, the proportional system fails to achieve the precision required at lower $\theta$ and $r$ and goes below the FIS output surface, and may not be able to generate significant $T$ at all. At the same time it is not able to produce large $T$ at large $r$ and $\theta$. It only approximates the sigmoid curve at moderate values of the inputs.

In Fig. 4.11 the continuous variation of $d$ and $\theta$ during game play and the generation of $T$ are explained. For clarity only 500 ms time span is chosen when the player was trying to turn right. The diameter $d$ is varying continuously and $\theta$ is continuously decreasing from $58°$ to $51°$. The skeleton is sampled at an interval of 50 ms (at a rate of 20 fps) and a simple sample and hold mechanism is used to store the sampled value.

**Table 4.2** Sample results for a particular instance for unknown subject

| Intended operation | $\theta$ | $d$ (normalized) | $\delta$ | $T$ for steering | Left/right | $T$ for accel./brake | Up/down |
|---|---|---|---|---|---|---|---|
| Turn Right | +42:3° | 0:68 | N.R. | 73:18 | Right | N.R. | N.R. |
| Turn left | −31.27° | 0.62 | N.R. | 45.92 | Left | N.R. | N.R. |
| Go Straight | +4.33° | 0.71 | N.R. | 5 | N.R. | N.R. | N.R. |
| Accelerate | N.R. | N.R. | 0.29 | N.R. | N.R. | 481 | Up |
| Brake | N.R. | N.R. | 0.23 | N.R. | N.R. | 162 | Down |

Straight Drive (frame number 128)          Turn Left (frame number 256)

Turn Right (frame number 314)

**Fig. 4.8** Game snapshots of Final Drive Fury

**Fig. 4.9** Output surface plot
for the proposed fuzzy
inference system

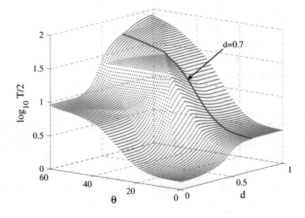

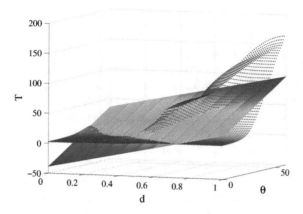

**Fig. 4.10** Comparison between FIS output and proportional output

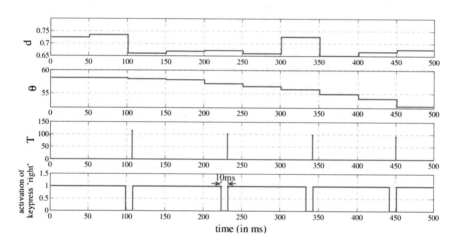

**Fig. 4.11** Sampling of $d$ and $\theta$ over time with generation of $T$

At the zero-th instant, the key-press duration $T$ is evaluated and the "key-down" event for right-arrow key is generated. $T$ is not evaluated further until the "key-release" event is actuated (after the time interval of current $T$) and one pulse of key-press is over. Then $T$ is again evaluated and this phenomenon goes on. Same is true for acceleration and braking where the "key-down" and "key-release" events for up-arrow and down-arrow keys are actuated in the same fashion. Between two key-press pulses, a guard-time of 10 ms is kept to avoid oversampling.

It can be observed from Fig. 4.11 that the output of the FIS actually relies on the pulse-width modulation of key-press pulse. According to the value of $d$ and $\theta$, the output $T$ modulates the time-duration of the virtual pressing of keys. That gives an effective method to control the turn as well as acceleration of the car.

## 4.6    Conclusion

The ideas we have proposed in this paper seek to provide a robust gesture controlled driving system that mimics the normal driving phenomena in the automo-
biles. We have sought to provide the most natural experience of driving while
playing a simulator racing game. The novelty of the proposed model taking natural
driving gestures as the input signals to a racing game is its less sensitivity to the
erroneous human perception and precise virtual key-press duration proportional to
the different quantities measured from the skeleton data. A Type-I Mamdani Fuzzy
Inference System is used to process the raw information received from MS Kinect
and the sigmoid nature of input-output relationship achieved by the FIS contributed
significantly to the ease of driving by precise control at lower angles and sharp turns
at higher angles, as well as incorporating both acceleration and brake for a complete
driving experience. The final output is taken in the form of key-board interrupt
signals with a certain key-press duration, which can be more sophistically integrated into the game engines at its development stage to provide a new gesture
driven control. The veracity of our claims is justified using the graphs that show the
relationship among random inputs and generated Key-press duration.

As this model is developed in accordance with the rigour and precision of a
simulator racer, the proposed approach can be readily applied to the similar type of
games, such as an "Arcade racer" that needs the activation of "Nitro" or
"Power-slide". Hidden Markov Models [3, 15] or Dynamic Time Warp [16] may be
used for recognition of some specific set of gestures along with the proposed here
ones for steering and acceleration.

Future work will focus on expanding the degree of freedom of the driving system
so that it can be applied to aircraft simulator where the imaginary steering wheel can
be replaced by an imaginary control stick. Again this work can be directly applied
to a robot-car where the output signal will be the pulse signal driving the
servo-motor.

## 4.7    Matlab Codes

```
% Code is written by Sriparna Saha
% Under the guidance of Prof. Amit Konar
dims=4;
joints=20;
fid = fopen('temp.txt');%input file
A = fscanf(fid, '%g', [1 inf]);
P=1;
total=length(A);
frames=total/(dims*joints);
F=zeros(frames,joints,dims);
```

```
for i=1:frames
  for j=1:joints
    for k=1:dims
      F(i,j,k)=A(P);
      P=P+1;
    end
  end
end
fclose(fid);
for iter=1:10:984%iter is for frame no
  figure(iter)
  hold on
  axis equal
  for i=1:joints
    scatter3(F(iter,i,1),F(iter,i,2),F(iter,i,3),'ok','filled');
  end
x=[ 4; 3; 3; 5; 6; 7; 9; 10; 11; 3; 2; 1; 1; 13; 14; 15; 17; 18; 19];
y=[ 3; 5; 9; 6; 7; 8; 10; 11; 12; 2; 1; 13; 17; 14; 15; 16; 18; 19; 20];
for i=1:length(x)
  plot3([F(iter,x(i),1) F(iter,y(i),1)],[F(iter,x(i),2)   F(iter,y
(i),2)],[F(iter,x(i),3) F(iter,y(i),3)],'Color','b','LineWidth',1.5);
end
end
```

**Results**: Sample run results are given in Fig. 4.1.

# References

1. Fraunhofer-Gesellschaft. "Gesture-driven Computers Will Take Computer Gaming To New Level." ScienceDaily. ScienceDaily, 7 Mar 2008
2. T.M. Bonanni, Person-tracking and gesture-driven interaction with a mobile robot using the Kinect sensor, Masters Thesis, Faculty of Engineering, Sapienza Universit'a Di Roma
3. L.E. Baum, T. Petrie, Statistical inference for probabilistic functions of finite state Markov Chains. Ann. Math. Stat. **37**(6), 15541563 (1966)
4. Y. Li, Hand gesture recognition using Kinect, Software Engineering and Service Science (ICSESS), 2012 3rd International Conference on IEEE, 22–24 June 2012, pp. 196–199
5. M. Oszust, M. Wysocki, Recognition of signed expressions observed by Kinect Sensor. Advanced Video and Signal Based Surveillance (AVSS), 2013 10th International Conference on IEEE, 2013, pp. 220–225
6. N. Burba, M. Bolas, D.M. Krum, E.A. Suma, Unobtrusive measurement of subtle nonverbal behaviors with the Microsoft Kinect. Virtual Reality Workshops (VR), 2012 IEEE, 2012, pp. 14
7. C. Zhang, Z. Bai, B. Cao, and J. Lin, Simulation and experiment of driving control system for electric vehicle. Int. J. Inf. Syst. Sci. **1**(3–4), 283–292

8.  M. Wada, Y. Kimura, Stability analysis of car driving with a joystick interface. IEEE 4th International Conference on Cognitive Infocommunications (CogInfoCom), 2–5 Dec 2013, pp. 493–496
9.  B, Pfleging, S. Schneegass, A. Schmidt, Multimodal interaction in the car-combining speech and gestures on the steering wheel. AutomotiveUI'12, October 17–19, Portsmouth, NH, USA
10. C. Sivertsen, Steering wheel input device having gesture recognition and angle compensation capabilities, Google Patents, 24 Jan 2013
11. M. Kato, Forza Motorsport 4 - Forza 4 Is A Finely Tuned Racing Machine, Game Informer, October 6, 2011
12. J. Solaro, "The Kinect Digital Out-of-Box Experience", Computer (Long. Beach. Calif)., pp. 9799, 2011
13. T. Leyvand, C. Meekhof, Y. C. Wei, J. Sun, and B. Guo, Kinect identity: technology and experience. Computer (Long. Beach. Calif). 44(4), 9496 (2011)
14. T. Dutta, Evaluation of the KinectTM sensor for 3-D kinematic measurement in the workplace. Appl. Ergon. 43(4), 645649 (2012)
15. L. Piyathilaka, S. Kodagoda, Gaussian mixture based HMM for human daily activity recognition using 3D skeleton features. Industrial Electronics and Applications (ICIEA), 2013 8th IEEE Conference on, pp. 567–572, 1921 June 2013
16. S. Masood, M. ParvezQureshi, M. B. Shah, S. Ashraf, Z. Halim, G. Abbas, Dynamic time wrapping based gesture recognition. Robotics and Emerging Allied Technologies in Engineering (iCREATE), 2014 International Conference on, pp. 205–210, 22–24 Apr 2014

# Chapter 5
# Type-2 Fuzzy Classifier Based Pathological Disorder Recognition

Gestures of a person symbolizing analogous pathological disorder are not always exclusive. Naturally, the gestural features of a subject suffering from the same pathological disorder vary widely over different instances. In presence of two or more gestural features, the variation of the attributes together makes the problem of pathological disorder recognition more convoluted. Wider variance in gestural features is the main source of uncertainty in the pathological disorder recognition, which has been addressed here using type-2 fuzzy sets (T2FSs). First, a type-2 fuzzy gesture space is created with the background knowledge of gestural features of different subjects for different pathological disorders. Second, the pathological disorder of an unknown gestural expression is recognized based on the consensus of the measured gestural features of the fuzzy gesture space. A fusion of interval and general type-2 Fuzzy Sets has been used to construct the fuzzy gestural space. An interval valued fuzzy set (IVFS) or interval type-2 fuzzy set (IT2FS) is constructed with primary membership functions for $M$ gestural features obtained from $N$ subjects, each having $L$ instances of gestural expressions for a given pathological disorder. The corresponding general type-2 fuzzy set (GT2FS) includes the secondary memberships for individual primary upper and lower membership curves in the respective IVFS. The secondary membership of a given source, characterizing the reliability in its primary membership assignment, is determined based on the amalgamated knowledge of all the subjects' primary membership functions. The representatives of each pathological disorder class are then evaluated by utilizing the attractive features of both IVFS and GT2FS. The pathological disorder of an unknown gestural expression is finally determined by identifying its minimal discrepancy with the existing class representatives. The adopted T2FS-based uncertainty management strategy, when applied to gender independent training and testing datasets, results in a classification accuracy of 93.41%.

Contributed by Pratyusha Rakshit, Sriparna Saha and Amit Konar

## 5.1  Introduction

Senile degeneration of bones, joints and soft tissues ensuing pathological disorder is of high importance for elderly health care. Aging cannot be halted but an early stage indication of the pathological disorder may help the doctors to pave a remedial way. It is observed that elderly patients are reluctant to consult doctors by visiting hospitals or clinics. Delayed detection of the disorders also makes the therapeutic procedure more prolonged and troublesome than usually. For detection of physical disabilities, the doctors mainly rely on a number of investigations including blood test, X-ray, magnetic resonance imaging and arthroscopy [1]. It is noteworthy that the arrangement of these tests at hospitals is essential for affirmation of the pathological disorders at matured stages. Simultaneously, recognition of the stage of the disorder requires placing the test instruments at home to take the general trend of behavior of aged patients into consideration. However, this seems to be unrealistic as the required instruments are costly and also not portable in many circumstances. This has motivated us to devise a system for autonomously identifying the pathological disorder of an unknown patient utilizing his/her gesture as the modality for the pathological disorder recognition at an early stage.

A gesture can be generated by any state or motion of a body or body parts. Gesture recognition aims at processing of the information that is ingrained only in the gesture and is not conveyed by verbal or written message. It encompasses two fundamental steps comprising feature extraction [2] and classification [3]. There has been extensive literature on autonomous gesture recognition over the last two decades. Among these works undertaken by Parajuli et al. [4] and Le et al. [5] on health monitoring using Kinect sensors [6], recognition of expression by Oszust and Wysocki [7] and analysis of children tantrum behavior by Yu et al. [8] need special mentioning. In [9–14], the authors proposed several computational approaches for gesture recognition, which have direct and indirect use in healthcare applications.

The present work aims at identifying sixteen pathological disorders from the gestural features of the patients having pain in different parts of the body. These gestural features are extracted from the twenty joint co-ordinates of the skeletons recorded by the Kinect sensor along with software development kit (SDK) [6]. However, these gestural features greatly depend on the physical states of the subjects. For example, gestural expressions taken from different instances of a subject suffering from the same pathological disorder, exhibit wide variation in individual features. Moreover, different subjects suffering from the same physical disability have diversities in their gestural features. A small but random variation of gestural features around specific fixed points is apparently exposed by the repetitive experiments with a large number of subjects, each enduring multiple instances of similar pathological disorder. The discrepancy between particular gestural expressions, taken from different instances of analogous pathological disorder of a specific subject can be treated as an intra-personal level uncertainty [15]. On the other hand, the variation in the given gestural expression of all individuals for similar pathological disorder can be regarded as inter-personal level uncertainty [16].

The uncertainty related to the variations in measured features of such ill-defined systems can be handled with classical *type-1 fuzzy sets* (T1FS) [17]. In T1FS, a single membership function (MF) is employed to signify the degree of uncertainty in measurements of a given gestural feature. For example, the fuzzy concept about *"joint angle is small"* is represented by a T1 MF, monotonically decreasing with *"joint angle"* that decays down to zero value at some angle, say 20°. Since a number of monotonically decreasing functions of *"joint angle"* can be constructed by each of the $N$ subjects to exemplify the T1 MF, recognizing the best functional curve to embody the given concept remains an unsolved problem. The above problem can be mitigated by considering the union of all potential T1 MFs describing the above fuzzy concept. This union is often referred to as *footprint of uncertainty* (FOU) in the terminology of *interval valued fuzzy set* (IVFS) [18] or *interval type-2 fuzzy sets* (IT2FSs) [19]. The FOU of an IVFS (or IT2FS) is a region bounded by the *upper* and *lower membership functions*, respectively denoted by UMF and LMF, representing the minimum and the maximum of the T1 MFS embedded in the FOU. The T1 MFs embedded in an FOU portrays inter-personal level uncertainty at each value of the gestural feature *"joint angle"* in the fuzzy context of *"joint angle is small"*, considering all $N$ experimental subjects together.

Nonetheless, the degree of correct assignment of primary membership values to all $N$ T1 MFs, embedded in the FOU is treated as unity in IVFS, which is not always correct. Over the last decade, the researches took initiatives [19, 20] to delineate the degree of the intra-personal uncertainty for each T1 MF embedded in an FOU. For each subject, the degree of certainty in appropriately assigning the primary membership *"joint angle is small"* at each distinct measurement of the given *"joint angle"* is represented by a two-dimensional MF. Intuitively, it is a function of the gestural feature *"joint angle"* and its primary memberships. The degree of certainty thus acquired for each distinct value of the measured *"joint angle"* and its corresponding primary membership for *"joint angle is small"* is entitled as the secondary membership grade. The resulting system encompassing $N$ primary MFs (obtained from $N$ subjects) with their corresponding secondary membership estimates at each distinct value of *"joint angle"* forms a *general type-2 fuzzy set* (GT2FS) [21] corresponding to the fuzzy concept *"joint angle is small"*.

Apparently, GT2FS is supposed to capture the uncertainty of a problem in a more realistic sense and hence provides better results in pathological disorder classification for its representational improvement over its IVFS counterpart. However, the performance of GT2FS is constrained by the users' inability to correctly indicate the secondary memberships. This chapter proposes a novel approach to evaluate the secondary membership of a given gestural feature $f$ based on the primary membership values of the UMF and LMF of the corresponding interval valued fuzzy gesture space of $f$. The proposed strategy assumes that the maximum and minimum values of the UMF and the LMF are assigned (by the experts) with more certainty. It has motivated us to assign the highest secondary membership grades to the measurements of the gestural feature $f$ where either a maximum or a minimum primary membership value has been attained by the UMF and the LMF of the given interval valued fuzzy gesture space of $f$. The secondary membership at other measurements of feature $f$ falls off with its distance away from the feature point corresponding to the extremal primary membership value of UMF and LMF respectively.

The chapter thus provides a novel approach to pathological disorder recognition from an unknown gestural expression by combining the attractive features of both IVFS and GT2FS. It employs a GT2 fuzzy gesture space, consisting of secondary membership functions of the UMF and the LMF of an IVFS, obtained from known gestural expressions of several patients suffering from various pathological disorders. The representatives of each pathological disorder class are afterwards evaluated using the composite opinion of the interval valued and generalized T2 fuzzy gesture spaces. The pathological disorder class of an unknown gestural expression is determined by computing the support of each class representative to the given gestural expression. The support is evaluated by considering an aggregation of secondary grades and the primary memberships of individual feature at the given measurement point. The class with the maximum support is selected. Experiments undertaken reveal that the classification accuracy of pathological disorder of an unknown person by the proposed stratagem is as high as 93.41%.

The chapter is divided into six sections. Section 5.2 provides fundamental definitions associated with Type-2 Fuzzy Sets (T2FSs). Section 5.3 deals with the principle of uncertainty management in fuzzy gesture space for pathological disorder recognition. Experimental details are given in Sect. 5.4 and the methods of performance analysis are undertaken in Sect. 5.5. Concluding points are listed in Sect. 5.6.

## 5.2   Preliminaries on Type-2 Fuzzy Sets

In this Section, we define a few terminologies on type-1 (T1) and type-2 (T2) fuzzy sets (FSs) [16, 18, 22–30], which are frequently used in the rest of the chapter. These definitions are referred to frequently throughout the chapter.

**Definition 1** A *type-1 fuzzy set A* defined on a universe of discourse $X$, is expressed as a two tuple [28], given by

$$A = \{(x, \mu_A(x)) | x \in X\} \tag{5.1}$$

Here $\mu_A(x)$ known as *type-1 membership function* (T1 MF) is a crisp number in [0, 1] for a generic element $x \in X$.

**Definition 2** A *general type-2 fuzzy set (GT2FS)* $\widetilde{A}$ [20] is described by (5.2).

$$\widetilde{A} = \{((x, u), \mu_{\widetilde{A}}(x, u)) | x \in X, u \in [0, 1]\} \tag{5.2}$$

Here $\mu_{\widetilde{A}}(x, u) \in [0, 1]$ is known as the *type-2 membership function* (T2 MF) corresponding to the generic element $x \in X$ with its primary membership $u \subseteq [0, 1]$. The support $J_x^u$ of $u$ for a generic element $x \in X$ is defined as follows.

$$J_x^u = [u|u \in [0, 1], \mu_{\widetilde{A}}(x, u) > 0] \tag{5.3}$$

**Definition 3** $\widetilde{A}$ is referred to as *interval valued fuzzy set* (IVFS) [18] or *interval type-2 fuzzy set* (IT2FS) [29, 31] provided

$$\mu_{\widetilde{A}}(x, u) = 1, \quad \text{for } x \in X, u \in [0, 1] \tag{5.4}$$

**Definition 4** The *footprint of uncertainty* (FOU) [16] is a bounded region representing the uncertainty in the primary memberships of an IVFS (or IT2FS) $\widetilde{A}$ and is defined for a fully connected IT2FS $\widetilde{A}$ [32] as

$$FOU(\widetilde{A}) = \cup_{x \in X} J_x^u \tag{5.5}$$

**Definition 5** The *lower* and *upper membership functions* (LMF and UMF) are T1 MFs, which represent the lower and upper bound of $FOU(\widetilde{A})$ and are respectively defined in (5.6) and (5.7) for a generic element $x \in X$.

$$\underline{\mu}_{\widetilde{A}}(x) = \inf\{u|u \in [0, 1], \mu_{\widetilde{A}}(x, u) > 0\} \tag{5.6}$$

$$\overline{\mu}_{\widetilde{A}}(x) = \sup\{u|u \in [0, 1], \mu_{\widetilde{A}}(x, u) > 0\} \tag{5.7}$$

It is evident from (5.3), (5.5), (5.6) and (5.7) that for a generic element $x \in X$

$$J_x^u = \{(x, u)|\forall u \in [\underline{\mu}_{\widetilde{A}}(x), \overline{\mu}_{\widetilde{A}}(x)]\} \tag{5.8}$$

**Definition 6** *Mendel-John Representation Theorem* (MJRT) [16]: According to MJRT, an IT2FS or IVFS $\widetilde{A}$ is the union of all T1 FSs embedded in $FOU(\widetilde{A})$. It signifies that it is possible to define operations on an IT2FS using T1FS mathematics [16, 33].

## 5.3  Uncertainty Management in Fuzzy Gesture Space for Pathological Disorder Recognition

Figure 5.1 provides a schematic architecture of the proposed strategy for pathological disorder recognition using the gestural expression of the subjects. Here, the measured gestural features are encoded using an amalgamation of IVFS (or IT2FS) and GT2FS. Accordingly, the uncertainty involved in correctly identifying the pathological disorder of an unknown gestural expression is regulated by the created

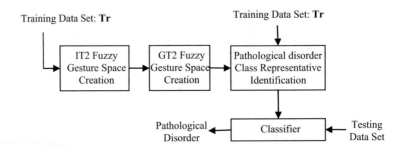

**Fig. 5.1** Block diagram for T2FS based pathological disorderrecognition

**Fig. 5.2** Pictorial
representation of the training
data sets for class $K_c$

| Class | Subjects $n$ | Instances $l$ | Features $f_{m,n}^{c,l}$ | | | |
|-------|-----|-----|-----|-----|-----|-----|
| | | | 1 | 2 | ........ | $M$ |
| $K_c$ | 1 | 1 | $f_{1,1}^{c,1}$ | $f_{2,1}^{c,1}$ | ........ | $f_{M,1}^{c,1}$ |
| | | 2 | $f_{1,1}^{c,2}$ | $f_{2,1}^{c,2}$ | ........ | $f_{M,1}^{c,2}$ |
| | | : | : | : | ........ | : |
| | | $L$ | $f_{1,1}^{c,L}$ | $f_{2,1}^{c,L}$ | ........ | $f_{M,1}^{c,L}$ |
| | 2 | 1 | $f_{1,2}^{c,1}$ | $f_{2,2}^{c,1}$ | ........ | $f_{M,2}^{c,1}$ |
| | | 2 | $f_{1,2}^{c,2}$ | $f_{2,2}^{c,2}$ | ........ | $f_{M,2}^{c,2}$ |
| | | : | : | : | ........ | : |
| | | $L$ | $f_{1,2}^{c,L}$ | $f_{2,2}^{c,L}$ | ........ | $f_{M,2}^{c,L}$ |
| | : | : | : | : | ........ | : |
| | | : | : | : | ........ | : |
| | $N$ | 1 | $f_{1,N}^{c,1}$ | $f_{2,N}^{c,1}$ | ........ | $f_{M,N}^{c,1}$ |
| | | 2 | $f_{1,N}^{c,2}$ | $f_{2,N}^{c,2}$ | ........ | $f_{M,N}^{c,2}$ |
| | | : | : | : | ........ | |
| | | $L$ | $f_{1,N}^{c,L}$ | $f_{2,N}^{c,L}$ | ........ | $f_{M,N}^{c,L}$ |

$$\mathbf{Tr}^c$$

fuzzy gesture space. The steps are elaborately demonstrated in the subsequent
sections. The proposed method involves a training data set **Tr**, which is pictorially
illustrated in Fig. 5.2, for a specific pathological disorder class $K_c$. The objective of
**Tr** is two-folds. First, it is used to generate the T2 fuzzy gestural feature space of
different pathological disorder classes. Second, it is used to determine the pathological disorder class representatives.

Let $F = \{f_1, f_2, \ldots, f_M\}$ be the set of $M$ gestural features and $K = \{ K_1, K_2, \ldots,$
$K_C\}$ be the $C$ classes of pathological disorder under question. The training data set
for a specific disorder class $K_c$, denoted by $\mathbf{Tr}^c$, is created by repeatedly measuring
a gestural feature $f_m$ for $L$ times on a particular subject, say $n$, experiencing the same
pathological disorder, say $K_c$. Let these feature samples be represented by $\{f_{m,n}^{c,l}\}$ for
$l = [1, L]$. This is done for all $M$ features of all $N$ subjects, all experiencing the same

pathological disorder $K_c$, i.e., for $m = [1, M]$ and $n = [1, N]$. It in turn results in $N \times L$, $M$-dimensional gestural feature data points in $\mathbf{Tr}^c$ (Fig. 5.2), denoted by

$$\overrightarrow{d}_n^{c,l} = [f_{1,n}^{c,l}, f_{2,n}^{c,l}, \ldots, f_{M,n}^{c,l}] \tag{5.9}$$

with $n = [1, N]$ and $l = [1, L]$. The same principle of generation of $\mathbf{Tr}^c$ is repeated for all pathological disorder classes $K_c$ with $c = [1, C]$ so that the resulting data set can be represented as

$$\mathbf{Tr} = [\mathbf{Tr}^1, \mathbf{Tr}^2, \ldots, \mathbf{Tr}^C]^T \tag{5.10}$$

### 5.3.1   Interval Valued or Interval Type-2 Fuzzy Space Formation for a Specific Gestural Feature of a Definite Class

The gestural feature data points of $\mathbf{Tr}^c$ are used to create the interval valued fuzzy space of a specific gestural feature $f_m$ corresponding to a particular pathological disorder class $K_c$. This is accomplished by following the principle of enhanced interval approach (EIA) proposed in [34, 35]. The important steps of EIA used in the present context are reproduced here.

1. **Identification of End Point Intervals**: Let $[a_n, b_n]$ be the end point intervals of the measurements of the given gestural feature $f_m$ for the $n$th subject obtained from $L$ instances of his/her gestural expressions for a specific pathological disorder class $K_c$. In other words,

$$a_n = \min_{l=1}^{L}\{f_{m,n}^{c,l}\} \tag{5.11}$$

and

$$b_n = \max_{l=1}^{L}\{f_{m,n}^{c,l}\} \tag{5.12}$$

The end point intervals are identified for the gestural expressions of all $N$ subjects of the data set $\mathbf{Tr}^c$ resulting in two sets of $N$ lower and $N$ upper end points, denoted by $\{a_n\}_{n=1}^{N}$ and $\{b_n\}_{n=1}^{N}$ respectively, corresponding to feature $f_m$ and pathological disorder class $K_c$.

2. **Outlier Processing**: This is done in two phases. In the first phase, the end point interval $[a_n, b_n]$ is judiciously selected if

$$a_n \in [Q_a(0.25) - 1.5IQR_a, Q_a(0.75) + 1.5IQR_a] \qquad (5.13)$$

and

$$b_n \in [Q_b(0.25) - 1.5IQR_b, Q_b(0.75) + 1.5IQR_b] \qquad (5.14)$$

where $Q_a(0.25)$ (or $Q_b(0.25)$), $Q_a(0.75)$ (or $Q_b(0.75)$) and $IQR_a$ (or $IQR_b$) are the first, third and inter-quartile range of the set of $N$ lower (or upper) end points $\{a^c_{m,n}\}^N_{n=1}$ (or $\{b^c_{m,n}\}^N_{n=1}$) respectively. This step thus reduces $N$ end point intervals into $N'$ end point intervals. Next, the interval length is computed for each of these $N'$ end point intervals, given as

$$L_n = b_n - a_n \qquad (5.15)$$

for $n = [1, N']$. The end point interval $[a_n, b_n]$ is kept if its respective interval length $L_n$ satisfies the following condition.

$$L_n \in [Q_L(0.25) - 1.5IQR_L, Q_L(0.75) + 1.5IQR_L] \qquad (5.16)$$

where $Q_L(0.25), Q_L(0.75)$ and $IQR_L$ are the first, third and inter-quartile range of the set of $N'$ interval lengths $\{L_n\}^{N'}_{n=1}$ respectively. This step thus further reduces the number of end point intervals from $N'$ to $N''$. Let $m_a$ (or $m_b$) and $\sigma_a$ (or $\sigma_b$) denote the sample mean and standard deviation of the set of $N''$ lower (or upper) end points $\{a_n\}^{N''}_{n=1}$ (or $\{b_n\}^{N''}_{n=1}$), survived after outlier processing, respectively.

3. **Tolerance Limit Processing**: This step presumes that the measurement data are normally distributed. It also comprises two phases. First the end point interval $[a_n, b_n]$ is selected for further processing if

$$a_n \in [m_a - k\sigma_a, m_a + k\sigma_a] \qquad (5.17)$$

and

$$b_n \in [m_b - k\sigma_b, m_b + k\sigma_b] \qquad (5.18)$$

This step thus reduces $N''$ end point intervals into $Y_1$ end point intervals. In (5.17) and (5.18), $k$ is determined from [36] for 95% confident assertion that the given limits consist of at least 95% of the measurement data. The resulting $Y_1$ end point intervals are further filtered by accepting only those intervals $[a_n, b_n]$ whose associated interval length $L_n$ satisfies

$$L_n \in [m_L - k'\sigma_L, m_L + k'\sigma_L] \tag{5.19}$$

Here $m_L$ and $\sigma_L$ denote the sample mean and standard deviation of the set of $Y_1$ interval lengths $\{L_n\}_{n=1}^{Y_1}$ respectively. In (5.19),

$$k' = \min(k_1, k_2, k_3) \tag{5.20}$$

where $k_1$ is determined in the same manner of calculating $k$ in (5.17) and (5.18) and

$$k_2 = m_L/\sigma_L \tag{5.21}$$

and

$$k_3 = (10 - m_L)/\sigma_L \tag{5.22}$$

This step thus results in $Y_2 < Y_1$ end point intervals. The sample means $m_a$ and $m_b$ and the standard deviations $\sigma_a$ and $\sigma_b$ of the lower and upper end points, $\{a_n\}_{n=1}^{Y_2}$ and $\{b_n\}_{n=1}^{Y_2}$, of the surviving intervals are respectively recomputed.

4. **Reasonable Interval Test**: This step is concerned with checking whether the intervals are overlapping. According to this test, the interval $[a_n, b_n]$ is accepted if

$$2m_a - \xi^* \leq a_n < \xi^* < b_n \leq 2m_b - \xi^* \tag{5.23}$$

where

$$\xi^* = \frac{\left(m_b\sigma_a^2 - m_a\sigma_b^2\right) \pm \sigma_a\sigma_b\left[(m_a - m_b)^2 + 2\left(\sigma_a^2 - \sigma_b^2\right)\log(\sigma_a/\sigma_b)\right]^{1/2}}{\sigma_a^2 - \sigma_b^2} \tag{5.24}$$

We thus finally end up with $N^* \leq Y_2$ end point intervals.

5. **FOU Selection**: This step is concerned with identifying the correct FOU for the set of end point intervals $\{[a_n, b_n]\}_{n=1}^{N^*}$, which have survived after outlier processing, tolerance limit processing and reasonable interval test. According to [34], the FOU of the set $\{[a_n, b_n]\}_{n=1}^{N^*}$ can be of three types, including interior, left shoulder and right shoulder. To accomplish this, first the sample means $m_a$ and $m_b$ and the standard deviations $\sigma_a$ and $\sigma_b$ of the lower and upper end points, $\{a_n\}_{n=1}^{N^*}$ and $\{b_n\}_{n=1}^{N^*}$, of the surviving $N^*$ intervals are respectively recomputed. We reproduce below the criteria for categorizing the set of end point intervals $\{[a_n, b_n]\}_{n=1}^{N^*}$ into one of these three FOUs for a specific gestural feature $f_m$ and pathological disorder class $K_c$.

Selection criteria for interior FOU:

$$m_b \leq 5.831 m_a - t_{\alpha, N_1^* - 1} \frac{\sigma_c}{\sqrt{N^*}}$$

$$m_b \leq 0.171 m_a + 8.29 - t_{\alpha, N_1^* - 1} \frac{\sigma_d}{\sqrt{N^*}} \qquad (5.25)$$

$$m_b \geq m_a$$

Selection criteria for left shoulder FOU:

$$m_b > 5.831 m_a - t_{\alpha, N_1^* - 1} \frac{\sigma_c}{\sqrt{N^*}}$$

$$m_b < 0.171 m_a + 8.29 - t_{\alpha, N_1^* - 1} \frac{\sigma_d}{\sqrt{N^*}} \qquad (5.26)$$

Selection criteria for right shoulder FOU:

$$m_b < 5.831 m_a - t_{\alpha, N^* - 1} \frac{\sigma_c}{\sqrt{N^*}}$$

$$m_b > 0.171 m_a + 8.29 - t_{\alpha, N^* - 1} \frac{\sigma_d}{\sqrt{N^*}} \qquad (5.27)$$

where $\sigma_c$ and $\sigma_d$ are the sample standard deviations of the sets $\{b_n - 5.83 a_n\}_{n=1}^{N^*}$ and $\{b_n - 0.171 a_n - 8.29\}_{n=1}^{N^*}$ respectively. Here $t_{\alpha, N^* - 1}$ represents the $t$-value for $(1 - \alpha)$ confidence level and $(N^* - 1)$ degree of freedom.

6. **FOU Parameter Estimation**: Once the FOU for the set of $N^*$ end point intervals $\{[a_n, b_n]\}_{n=1}^{N^*}$ is correctly identified among the three possible FOUs corresponding to a specific pair of gestural feature $f_m$ and pathological disorder class $K_c$, the lower and upper end points, $a_n$ and $b_n$, of each interval are transformed into the parameters $a_{MF,n}$ and $b_{MF,n}$ of a T1FS respectively for $n = [1, N^*]$ as given below in Table 5.1.

   In Table 5.1, $\mu_{A_n}(f_m)$ represents the primary membership of the gestural feature $f_m$ belonging to the pathological disorder class $K_c$ obtained from the experiments on the $n$th subject. The resulting T1FSs (obtained from experiments on $N^*$ satisfies) are referred to as embedded T1FSs, denoted by $\{A_n\}_{n=1}^{N^*}$, corresponding to the gestural feature $f_m$ and the pathological disorder class $K_c$.

7. Eventually, the set of $N^*$ T1 MFs for a given feature $f_m$ taken from $N$ subjects (with $N^* < N$ due to pre-processing and filtering of end point intervals using outlier processing, tolerance limit processing and reasonable interval test) together form an IVFS, $\tilde{A}$ for the pathological disorder class $K_c$ corresponding to

**Table 5.1** Transformation of the uniformly distributed end point interval $[a_n, b_n]$ into the parameters $a_{\mathrm{MF},n}$ and $b_{\mathrm{MF},n}$ of a T2FS

| Name | Membership function | Transformation |
|---|---|---|
| Interior/ symmetric triangle | | $a_{\mathrm{MF},n} = \dfrac{1}{2}[(a_n + b_n) - \sqrt{2}(b_n - a_n)]$ <br><br> $b_{\mathrm{MF},n} = \dfrac{1}{2}[(a_n + b_n) + \sqrt{2}(b_n - a_n)]$ |
| Left shoulder | | $a_{\mathrm{MF},n} = \dfrac{(a_n + b_n)}{2} - \dfrac{(b_n - a_n)}{\sqrt{6}}$ <br><br> $b_{\mathrm{MF},n} = \dfrac{(a_n + b_n)}{2} + \dfrac{\sqrt{6}(b_n - a_n)}{3}$ |
| Right shoulder | | $a'_n = M - b_n$ <br> $b'_n = M - a_n$ <br><br> $a_{\mathrm{MF},n} = M - \dfrac{(a'_n + b'_n)}{2} - \dfrac{\sqrt{6}(b'_n - a'_n)}{3}$ <br><br> $b_{\mathrm{MF},n} = M - \dfrac{(a'_n + b'_n)}{2} + \dfrac{(b'_n - a'_n)}{\sqrt{6}}$ |

the gestural feature $f_m$ as shown in Fig. 5.3. Following the representation theorem of Definition 6 [16], $\widetilde{A}$ can be symbolically defined by

$$\widetilde{A} = \bigcup_{n=1}^{N^*} A_n \tag{5.28}$$

where

$$A_n = \{(f_m, \mu_{A_n}(f_m)) | \forall f_m\} \tag{5.29}$$

It is apparent that FOU of $\widetilde{A}$, obtained from the mathematical models of UMF $\overline{\mu}_{\widetilde{A}}(f_m)$ and LMF $\underline{\mu}_{\widetilde{A}}(f_m)$ of [34], can be of three types, including interior, left shoulder and right shoulder based on the criteria given in (5.25) to (5.27). It also confirms that the resulting $N^*$ embedded TI FSs, $\{A_n\}_{n=1}^{N^*}$, are completely contained within the FOU of $\widetilde{A}$.

Eventually, both intra- and inter-personal level uncertainty are now induced in the measurement attribute $f_m$. The intra-level uncertainty accounts for the pre-assumption of a specific form (for example, interior or left shoulder or right shoulder) of primary or T1 MFs, and the inter-level uncertainty occurs due to multiplicity of the MFs for $N^*$ subjects. The uncertainty involved in the present

**Fig. 5.3** IVFS formation
from primary memberships of
gestural feature $f_m$ belonging
to pathological disorder class
$K_c$ satisfying three conditions

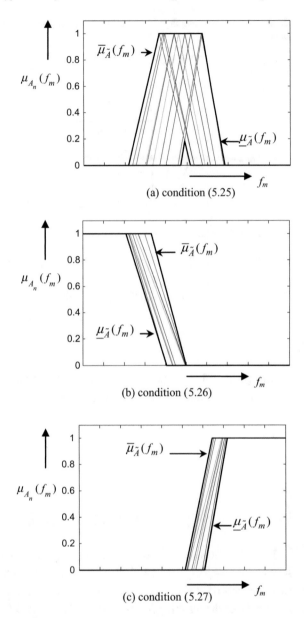

problem has been addressed here by T2FS. It handles intra- and inter-personal level
uncertainty compositely by blending the primary and the secondary membership
functions.

The outline of generating interval valued fuzzy gesture space of gestural feature
$f_m$ belonging to the pathological disorder class $K_c$ is given in Fig. 5.4 for $m \in [1, M]$
and $c \in [1, C]$.

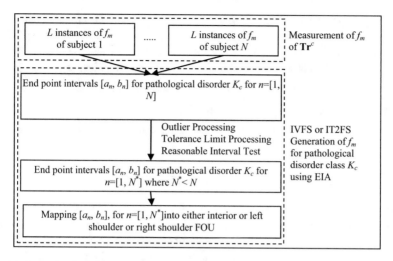

**Fig. 5.4**  Interval valued fuzzy space creation of feature $f_m$ for class $K_c$

## 5.3.2   General Type-2 Fuzzy Space Creation for a Specific Gestural Feature of a Definite Class

This section proposes a novel approachwhich reduces the degree of intra-level uncertainty incurred in UMF $\overline{\mu}_{\widetilde{A}}(f_m)$ and LMF $\underline{\mu}_{\widetilde{A}}(f_m)$ (due to the primary membership assignment of $\mu_{A_n}(f_m)$ for a given gestural feature $f_m$ acquired from the $L$ postural instances of the $n$th subject with pathological disorder $K_c$ for $n = [1, N^*]$), using their corresponding secondary memberships $\mu_{\widetilde{U}}(f_m, \overline{\mu}_{\widetilde{A}}(f_m))$ and $\mu_{\widetilde{L}}(f_m, \underline{\mu}_{\widetilde{A}}(f_m))$ respectively. Consequently, a special kind of GT2FS $\widetilde{A}$ is constructed for gesture recognition which has only two embedded T2 FSs, including $\widetilde{U}$ and $\widetilde{L}$, such that

$$\widetilde{U} = \{((f_m, \overline{\mu}_{\widetilde{A}}(f_m)), \mu_{\widetilde{U}}(f_m, \overline{\mu}_{\widetilde{A}}(f_m))) | \forall f_m\} \tag{5.30}$$

and

$$\widetilde{L} = \{((f_m, \underline{\mu}_{\widetilde{A}}(f_m)), \mu_{\widetilde{L}}(f_m, \underline{\mu}_{\widetilde{A}}(f_m))) | \forall f_m\} \tag{5.31}$$

The reduction in the degree of uncertainty facilitates the upgrading in the classification accuracy of pathological disorder detection at the cost of additional run-time complexity required to evaluate T2 secondary distributions. Despite its robustness in alleviating the uncertainty, the applications of GT2FS remains confined over the last two decades because of the users' inadequate knowledge regarding appropriate secondary membership assignment. The problem of manual assignment of secondary memberships here is circumvented by extracting T2 MF from its T1 counterpart using two stratagems. A brief outline to the adopted strategies for evaluating the secondary membership function $\mu_{\tilde{U}}(f_m, \overline{\mu}_{\tilde{A}}(f_m))$ is given in this section, with $\overline{\mu}_{\tilde{A}}^{max}$ and $\overline{\mu}_{\tilde{A}}^{min}$ symbolizing the maximum and the minimum primary membership values of the UMF $\overline{\mu}_{\tilde{A}}(f_m).\mu_{\tilde{L}}(f_m, \underline{\mu}_{\tilde{A}}(f_m))$ is evaluated following the same approach and hence is not explained in details for space economy.

### Strategy-1: Secondary membership assignment corresponding to maxima and minima of UMF

Consider a problem of grading of students in a course. Let the marks obtained by a student be a random variable $v$ in [0, 100]. It is apparent from common sense that in case a student obtains 0 or 100, the certainty (membership in the context of fuzzy sets) of his obtaining 0 or 100 is very high irrespective of the examiners. Here, the primary membership of UMF being a variable in $[\overline{\mu}_{\tilde{A}}^{min}, \overline{\mu}_{\tilde{A}}^{max}]$, the secondary membership, defining the degree of correctness in primary assignment, is expected to be high (close to 1) [19, 37] for the occurrence of the primary membership at the boundaries of the above interval. Thus, formally secondary membership

$$\mu_{\tilde{U}}(f_m, \overline{\mu}_{\tilde{A}}(f_m)) = 1 \tag{5.32}$$

for all values of $f_m$ where $\overline{\mu}_{\tilde{A}}(f_m) = \overline{\mu}_{\tilde{A}}^{max}$ and $\overline{\mu}_{\tilde{A}}(f_m) = \overline{\mu}_{\tilde{A}}^{min}$ if $\overline{\mu}_{\tilde{A}}(f_m) = \overline{\mu}_{\tilde{A}}^{max}$ or $\overline{\mu}_{\tilde{A}}(f_m) = \overline{\mu}_{\tilde{A}}^{min}$ and all such feature points (available in the entire span of feature space) are included in set $F$. In other words,

$$F = \{f_m | \mu_{\tilde{U}}(f_m, \overline{\mu}_{\tilde{A}}(f_m)) = 1\} \tag{5.33}$$

Expression (5.32) is also valid for LMF $\underline{\mu}_{\tilde{A}}(f_m)$. However, there may not exist enough background knowledge to appropriately assign the secondary membership grades corresponding to the non-optimal primary membership values of $\overline{\mu}_{\tilde{A}}(f_m)$ where $\overline{\mu}_{\tilde{A}}^{min} < \overline{\mu}_{\tilde{A}}(f_m) < \overline{\mu}_{\tilde{A}}^{max}$. These are evaluated using the second strategy.

### Strategy-2: Secondary membership assignment corresponding to non-optimal (neither maxima nor minima) primary memberships of UMF

Let $f_m$ be a gestural feature where primary membership value of $\overline{\mu}_{\tilde{A}}(f_m)$ is neither maximum nor minimum. Naturally, the secondary membership $\mu_{\tilde{U}}(f_m, \overline{\mu}_{\tilde{A}}(f_m))$ will be smaller than 1. Again let $\hat{f}_m \in F$ be the nearest feature point to $f_m$ where the primary membership value of $\overline{\mu}_{\tilde{A}}(\hat{f}_m)$ is either $\overline{\mu}_{\tilde{A}}(\hat{f}_m) = \overline{\mu}_{\tilde{A}}^{max}$ or $\overline{\mu}_{\tilde{A}}(\hat{f}_m) = \overline{\mu}_{\tilde{A}}^{min}$ (i.e., $\mu_{\tilde{U}}(\hat{f}_m, \overline{\mu}_{\tilde{A}}(\hat{f}_m)) = 1$). Symbolically,

$$\left|\hat{f}_m - f_m\right| = \arg\left(\min_{\forall f_l \in F}(\left|f_l - f_m\right|)\right) \qquad (5.34)$$

In this chapter, an exponential fall-off in $\mu_{\widetilde{U}}(f_m, \overline{\mu}_{\widetilde{A}}(f_m))$ at the given gestural feature value $f_m$ away from $\mu_{\widetilde{U}}(\hat{f}_m, \overline{\mu}_{\widetilde{A}}(\hat{f}_m)) = 1$ is modeled as given below [19].

$$\begin{aligned} \mu_{\widetilde{U}}(f_m, \overline{\mu}_{\widetilde{A}}(f_m)) &= \mu_{\widetilde{U}}(\hat{f}_m, \overline{\mu}_{\widetilde{A}}(\hat{f}_m)) \times \exp(-\alpha \times |\hat{f}_m - f_m|) \\ &= 1 \times \exp(-\alpha \times |\hat{f}_m - f_m|) \end{aligned} \qquad (5.35)$$

Expression (5.35) is devised on the basis of the intuition that the degree of certainty in assigning non-optimal primary membership of $\overline{\mu}_{\widetilde{A}}(f_m)$ at a gestural feature measurement $f_m$ diminishes with its distance away from the nearest feature point $\hat{f}_m$ where the primary membership value of $\overline{\mu}_{\widetilde{A}}(\hat{f}_m)$ is either maximum or minimum. In other words, a small value of $|\hat{f}_m - f_m|$ signifies higher degree of accuracy in assigning the primary membership of $\overline{\mu}_{\widetilde{A}}(f_m)$ at $f_m$. Contrarily, a lower grade of certainty in the assignment of the primary membership of $\overline{\mu}_{\widetilde{A}}(f_m)$ is indicated by an estimated large value of $|\hat{f}_m - f_m|$. Expression (5.34) and (5.35) hold for all distinct sampled values of $f_m$ where $\overline{\mu}_{\widetilde{A}}^{\min} < \overline{\mu}_{\widetilde{A}}(f_m) < \overline{\mu}_{\widetilde{A}}^{\max}$. Here $\alpha$ is the decay rate.

The decay rate $\alpha$ is estimated from a predefined value of minimum secondary membership $\mu_{\widetilde{U}}^{\min}$ of the UMF by the following procedure. The feature point $f_c$ is identified which is farthest from its nearest feature point $\hat{f}_c \in F$ (i.e., $\mu_{\widetilde{U}}(\hat{f}_c, \overline{\mu}_{\widetilde{A}}(\hat{f}_c)) = 1$). Symbolically,

$$\left|\hat{f}_c - f_c\right| = \arg\left(\max_{\forall f_l \in F, \forall f_m \notin F}(\left|f_l - f_m\right|)\right) \qquad (5.36)$$

It is apparent from (5.35) and (5.36) that $\mu_{\widetilde{U}}(f_c, \overline{\mu}_{\widetilde{A}}(f_c))$ should be minimum and hence, is set as

$$\mu_{\widetilde{U}}(f_c, \overline{\mu}_{\widetilde{A}}(f_c)) = \exp(-\alpha \times |\hat{f}_c - f_c|) = \mu_{\widetilde{U}}^{\min} \Rightarrow \alpha = \frac{-\ln\left(\mu_{\widetilde{U}}^{\min}\right)}{|\hat{f}_c - f_c|} \qquad (5.37)$$

Evidently, $\alpha$ can be evaluated from the known predefined value of $\mu_{\widetilde{U}}^{\min}$ and the evaluated value of $|\hat{f}_c - f_c|$.

A pseudo-code to compute $\mu_{\widetilde{U}}(f_m, \overline{\mu}_{\widetilde{A}}(f_m))$ from its primary counterpart is given below. It is mention worthy that $\mu_{\widetilde{L}}(f_m, \underline{\mu}_{\widetilde{A}}(f_m))$ can be obtained by adopting the same pseudo code used for generating $\mu_{\widetilde{U}}(f_m, \overline{\mu}_{\widetilde{A}}(f_m))$. While evaluating

$\mu_{\tilde{L}}(f_m, \mu_{\tilde{A}}(f_m))$, the input to the pseudo-code will be $\underline{\mu}_{\tilde{A}}(f_m)$ instead of $\overline{\mu}_{\tilde{A}}(f_m)$. The remaining procedure will remain the same.

***Pseudo Code***:

---

**Procedure evaluate_secondary_UMF**

**Input:** Upper membership function $\overline{\mu}_{\tilde{A}}(f_m)$ of the gestural feature $f_m$ belonging to the pathological disorder class $K_c$, $R$ measured sample points of the gestural feature $f_m$, $[f_{m1}, f_{mR}]$ such that $f_{m1} < f_{m2} < ... < f_{mR}$.

**Output:** Secondary membership distribution $\mu_{\tilde{U}}(f_m, \overline{\mu}_{\tilde{A}}(f_m))$ corresponding to $\overline{\mu}_{\tilde{A}}(f_m)$.

---

**Begin**

     1.     Obtain $\overline{\mu}_{\tilde{A}}^{min}$ and $\overline{\mu}_{\tilde{A}}^{max}$ from $\overline{\mu}_{\tilde{A}}(f_m)$.

     2.     Set $F \leftarrow NULL$.

     3.     **For** $f_m = f_{m1}$ to $f_m = f_{mR}$ **do**

     **Begin**

         **If** $\overline{\mu}_{\tilde{A}}(f_m) = \overline{\mu}_{\tilde{A}}^{min}$ or $\overline{\mu}_{\tilde{A}}(f_m) = \overline{\mu}_{\tilde{A}}^{max}$ **then do**

         **Begin**

             Set $F \leftarrow F \cup \{f_m\}$.

             Set $\mu_{\tilde{U}}(f_m, \overline{\mu}_{\tilde{A}}(f_m)) \leftarrow 1$.

         **End If.**

     **End For.**

     4.     Identify $\hat{f}_m \in F$ closest to $f_m$, $\forall \, f_m \in [f_{m1}, f_{mR}]$ but $\forall f_m \notin F$ and thus evaluate $| \hat{f}_m - f_m |$.

     5.     Identify $| \hat{f}_c - f_c | = \arg\left( \displaystyle\max_{f_m = f_{m1}, f_m \notin F}^{f_{mR}} (| \hat{f}_m - f_m |) \right)$.

     6.     Obtain $\alpha$ from (37).

     7.     Obtain $\mu_{\tilde{U}}(f_m, \overline{\mu}_{\tilde{A}}(f_m)) = \exp(-\alpha \times | \hat{f}_m - f_m |) \; \forall \; f_m \in [f_{m1}, f_{mR}]$ but $\forall f_m \notin F$.

**End.**

---

Figure 5.5 pictorially exemplifies the evaluation of secondary membership grade $\mu_{\tilde{U}}(f_m, \overline{\mu}_{\tilde{A}}(f_m))$ for a trapezoidal $\overline{\mu}_{\tilde{A}}(f_m)$ for a span of the feature points $[f_1, f_9]$. We have intentionally taken $f_3 = (f_1 + f_5)/2$ and $f_8 = (f_7 + f_9)/2$.

## Application of Strategy-1

It is evident from Fig. 5.5 that $\overline{\mu}_{\tilde{A}}(f_5) = \overline{\mu}_{\tilde{A}}(f_6) = \overline{\mu}_{\tilde{A}}(f_7) = \overline{\mu}_{\tilde{A}}^{max}$ and $\overline{\mu}_{\tilde{A}}(f_1) = \overline{\mu}_{\tilde{A}}(f_9) = \overline{\mu}_{\tilde{A}}^{min}$. Hence from (5.32) and (5.33), we obtain

$$\mu_{\tilde{U}}(f_m, \overline{\mu}_{\tilde{A}}(f_m)) = 1 \quad \text{for } f_m \in F \tag{5.38}$$

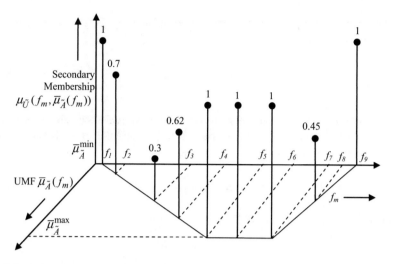

**Fig. 5.5** Assignment of secondary membership for UMF

where

$$F = \{f_1, f_5, f_6, f_7, f_9\} \tag{5.39}$$

**Application of Strategy-2**

This step is concerned with first identifying the nearest feature point $\hat{f}_m \in F$ corresponding to each feature point $f_m \in [f_1, f_9]$ but $f_m \notin F$ using (5.34). This results in

$$\left.\begin{array}{l} \hat{f}_2 = f_1 \\ \hat{f}_3 = \text{either } f_1 \text{ or } f_5 \\ \hat{f}_4 = f_5 \\ \hat{f}_8 = \text{either } f_7 \text{ or } f_9 \end{array}\right\} \tag{5.40}$$

The absolute differences $\left|\hat{f}_m - f_m\right|$ for $f_m = \{f_2, f_3, f_4, f_8\}$ are then sorted in the ascending order, which yields the sorted list as

$$\left\{\left|\hat{f}_2 - f_2\right|, \left|\hat{f}_4 - f_4\right|, \left|\hat{f}_8 - f_8\right|, \left|\hat{f}_3 - f_3\right|\right\} \tag{5.41}$$

From (5.36) and (5.41), we obtain $f_c = f_3$ which is the feature point farthest from $\hat{f}_c \in F$. Thus we set

$$\mu_{\tilde{U}}(f_3, \overline{\mu}_{\tilde{A}}(f_3)) = \exp(-\alpha \times \left|\hat{f}_3 - f_3\right|) = \mu_{\tilde{U}}^{min} = 0.3 \text{ (predefined)} \tag{5.42}$$

The decay rate $\alpha$ is then evaluated from (5.42) using $\hat{f}_3 =$ either $f_1$ or $f_5$ as $f_3 = (f_1 + f_5)/2$, given as

$$\alpha = -\frac{\ln(0.3)}{\left|\hat{f}_3 - f_3\right|} \tag{5.43}$$

The secondary memberships at $f_2, f_4$ and $f_8$ are next evaluated using (35) and the $\alpha$ value obtained from (5.43). It is evident from (35) and the sorted list in (41) that since $\left|\hat{f}_2 - f_2\right| < \left|\hat{f}_4 - f_4\right| < \left|\hat{f}_8 - f_8\right|$, therefore,

$$\mu_{\widetilde{U}}(f_2, \overline{\mu}_{\widetilde{A}}(f_2)) > \mu_{\widetilde{U}}(f_4, \overline{\mu}_{\widetilde{A}}(f_4)) > \mu_{\widetilde{U}}(f_8, \overline{\mu}_{\widetilde{A}}(f_8)) \tag{5.44}$$

Application of (5.35) and (5.43) yield

$$\left.\begin{array}{l} \mu_{\widetilde{U}}(f_2, \overline{\mu}_{\widetilde{A}}(f_2)) = 0.7 \\ \mu_{\widetilde{U}}(f_4, \overline{\mu}_{\widetilde{A}}(f_4)) = 0.62 \\ \mu_{\widetilde{U}}(f_8, \overline{\mu}_{\widetilde{A}}(f_8)) = 0.45 \end{array}\right\} \tag{5.45}$$

Figure 5.6 pictorially represents the generation of secondary memberships of UMF and LMF for three possible FOUs. Figure 5.6a.2–c.2 (or Fig. 5.6a.4–c.4) pictorially represent the two-dimensional view of secondary membership (SM) of UMF (or SM of LMF) with the sampled value of the gestural feature $f_m$. It is evident from Fig. 5.6a.1–c.1 and a.2–c.2 (or Fig. 5.6a.1–c.1 and a.4–c.4) that SM of UMF (or SM of LMF) attains the highest value of unity at the gestural feature value $f_m$ where the primary membership (PM) value of UMF (or PM of LMF) is either maximum or minimum. It indicates the maximum certainty in assigning the PM value to the UMF (or LMF). However, for other feature points $f_m$, SM of UMF(or SM of LMF) exponentially decreases with an increase in the distance from its nearest feature point $\hat{f}_m$, where the PM value of UMF (or LMF) is extremum. Figure 5.6a.3–c.3 (or Fig. 5.6a.5–c.5) pictorially represent the three-dimensional view comprising the PM and SM of UMF (or LMF) for different sampled values of gestural feature points. The outline of generating GT2 fuzzy gesture space of gestural feature $f_m$ belonging to the pathological disorder class $K_c$ is given in Fig. 5.7 for $m \in [1, M]$ and $c \in [1, C]$.

### 5.3.3  Evaluating Degree of Membership of a Specific Gestural Feature to a Definite Class

After obtaining $\mu_{\widetilde{U}}(f_m, \overline{\mu}_{\widetilde{A}}(f_m))$ and $\mu_{\widetilde{L}}(f_m, \underline{\mu}_{\widetilde{A}}(f_m))$, both embedded in $\widetilde{A}$, the degree of membership $Q_m^c$ of the gestural feature $f_m$ to belong to the pathological disorder class $K_c$ is determined. This is accomplished by utilizing a composite measure of the

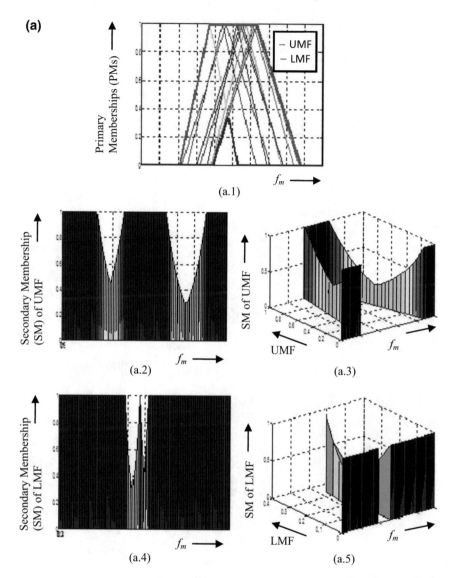

**Fig. 5.6   a** Representation of UMF and LMF for interior FOU. **b** Representation of UMF and LMF for left shoulder FOU. **c** Representation of UMF and LMF for right shoulder FOU

membership of the gestural feature data points $f_{m,n}^{c,l}$ of $\mathbf{Tr}^c$ for belonging to the pathological disorder class $K_c$ for $n = [1, N^*]$ and $l = [1, L]$. It is worth mentioning that $N^* \times L$ data points of $\mathbf{Tr}^c$ are sustained out of $N \times L$ data points for the specific feature $f_m$ after the pre-processing steps of EIA approach. To obtain $Q_m^c$, first an aggregated measure of the gestural feature $f_m$ of the $n$th subject of $\mathbf{Tr}^c$ is obtained by consulting its corresponding $L$ instances. The aggregated measure is given by

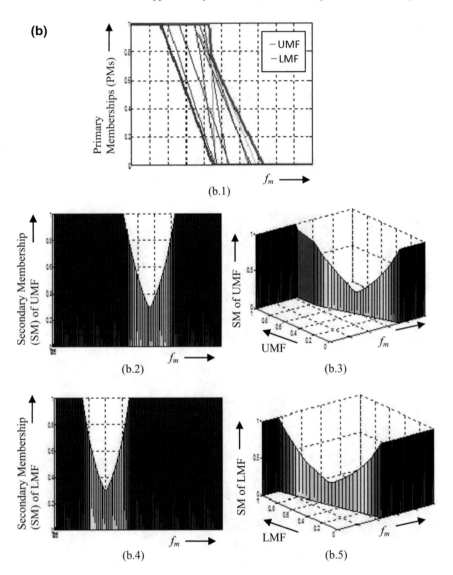

**Fig. 5.6** (continued)

$$f_{m,n}^c = \frac{\sum_{l=1}^{L} f_{m,n}^{c,l}}{L} \qquad (5.46)$$

It is evaluated for all subjects $n = [1, N^*]$. It is evident from Fig. 5.8 that the extent of belongingness of the given gestural feature $f_{m,n}^c$ to the pathological disorder class $K_c$ is validated by the UMF and LMF to a degree of $\overline{\mu}_{\widetilde{A}}(f_{m,n}^c)$ and

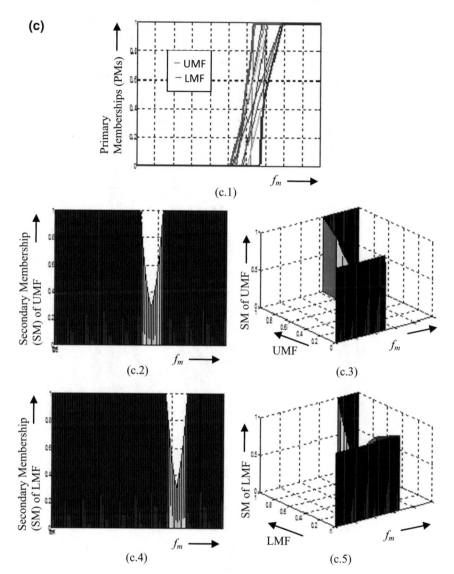

**Fig. 5.6**   (continued)

$\underline{\mu}_{\widetilde{A}}(f_{m,n}^c)$ with certainties $\mu_{\widetilde{U}}\left(f_{m,n}^c, \overline{\mu}_{\widetilde{A}}(f_{m,n}^c)\right)$ and $\mu_{\widetilde{L}}\left(f_{m,n}^c, \underline{\mu}_{\widetilde{A}}(f_{m,n}^c)\right)$ respectively. The extent of belongingness of the given gestural feature $f_{m,n}^c$ to the pathological disorder class $K_c$, induced by the combined conformity of UMF and LMF of $\widetilde{A}$ can be computed as given below.

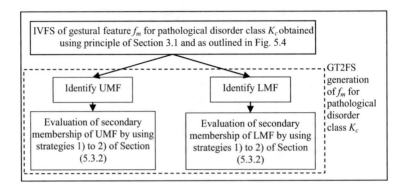

**Fig. 5.7** GT2 fuzzy space creation of feature $f_m$ for class $K_c$

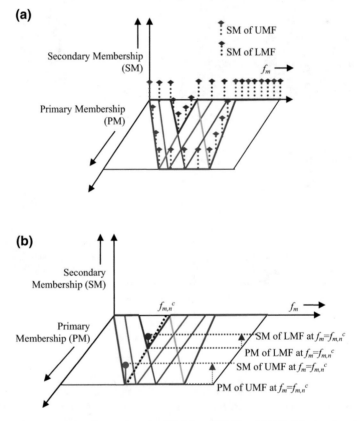

**Fig. 5.8** **a** Representation of secondary memberships of UMF and LMF of feature $f_m$ to belong to class $K_c$. **b** Representation of degree of membership of $f_{m,n}^c$ to belong to class $K_c$ as supported by UMF and LMF

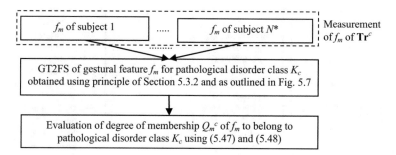

**Fig. 5.9**  Evaluation of degree of membership of feature $f_m$ to belongs to class $K_c$

$$Q_{m,n}^c = \frac{\overline{\mu}_{\tilde{A}}(f_{m,n}^c) \times \mu_{\tilde{U}}(f_{m,n}^c, \overline{\mu}_{\tilde{A}}(f_{m,n}^c)) + \underline{\mu}_{\tilde{A}}(f_{m,n}^c) \times \mu_{\tilde{L}}(f_{m,n}^c, \underline{\mu}_{\tilde{A}}(f_{m,n}^c))}{\mu_{\tilde{U}}(f_{m,n}^c, \overline{\mu}_{\tilde{A}}(f_{m,n}^c)) + \mu_{\tilde{L}}(f_{m,n}^c, \underline{\mu}_{\tilde{A}}(f_{m,n}^c))} \tag{5.47}$$

Expression (5.47) is evaluated for all feature data points of $\mathbf{Tr}^c$, i.e., for $n = [1, N^*]$, corresponding to the gestural feature $f_m$. Considering the composite opinion of $\left\{Q_{m,n}^c\right\}_{n=1}^{N^*}$, the effective primary membership of the gestural feature $f_m$ to belong to pathological disorder class $K_c$ can be obtained as

$$Q_m^c = \frac{\sum_{n=1}^{N^*} Q_{m,n}^c}{N^*} \tag{5.48}$$

The outline of evaluating the degree of membership $Q_m^c$ of gestural feature $f_m$ to belong to the pathological disorder class $K_c$ is given in Fig. 5.9 for $m \in [1, M]$ and $c \in [1, C]$.

### 5.3.4  Identifying Representative of a Definite Class

Expression (5.48) is evaluated all features $m = [1, M]$ following the principle in Sect. 5.3.3. Eventually, the pathological disorder class $K_c$ can be represented by an $M$-dimensional vector, denoted as

$$\overrightarrow{Q}^c = [Q_1^c, Q_2^c, \ldots, Q_M^c] \tag{5.49}$$

The vector is referred to as the class representative of pathological disorder $K_c$. The outline of identifying the representative of the pathological disorder class $K_c$ using the degree of membership of all gestural feature $f_m$ to belong to the pathological disorder class $K_c$ is given in Fig. 5.10 for $m = [1, M]$ and $c \in [1, C]$.

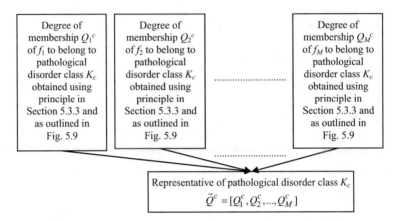

**Fig. 5.10** Identification of representative of class $K_c$

## 5.3.5  Identifying Representative of All Classes

Repeating step 5.3.4 for all variants of pathological disorders will ultimately yield $C$, $M$-dimensional class representatives

$$\overrightarrow{Q}^c = [Q_1^c, Q_2^c, \ldots, Q_M^c] \quad \text{for } c = [1, C] \tag{5.50}$$

Each $\overrightarrow{Q}^c$ corresponds to one pathological disorder class $K_c$ for $c = [1, C]$.

## 5.3.6  Principles Used to Identify Pathological Disorder Class from an Unknown Gestural Posture

Let the set of $M$ gestural features extracted from an anonymous gestural posture be $F' = \{f_1', f_2', \ldots, f_M'\}$. Then implementing the similar stratagem discussed in Sect. 5.3.3, the primary membership of the gestural feature $f_m'$ to the pathological disorder class $K_c$, induced by the mutual approval of UMF and LMF (of IVFS) corresponding to feature $f_m$ and class $K_c$ can be computed as given below for $m \in [1, M]$ and $c \in [1, C]$.

$$Q_m^{\prime c} = \frac{\overline{\mu}_{\tilde{A}}(f_m') \times \mu_{\tilde{U}}(f_m', \overline{\mu}_{\tilde{A}}(f_m')) + \underline{\mu}_{\tilde{A}}(f_m') \times \mu_{\tilde{L}}(f_m', \underline{\mu}_{\tilde{A}}(f_m'))}{\mu_{\tilde{U}}(f_m', \overline{\mu}_{\tilde{A}}(f_m')) + \mu_{\tilde{L}}(f_m', \underline{\mu}_{\tilde{A}}(f_m'))} \tag{5.51}$$

It is evident from the pseudo-code of **evaluate_secondary_UMF** (under Sect. 5.3.2) that the PM $\overline{\mu}_{\tilde{A}}(f_m)$ and SM $\mu_{\tilde{U}}(f_m, \overline{\mu}_{\tilde{A}}(f_m))$ of UMF are obtained for a specific set of $R$ sampled values of the gestural feature $f_m$, given by $\{f_{m1}, f_{m2}, \ldots, f_{mR}\}$. Hence the extracted feature $f_m'$ of the unknown gesture may not exactly match with any existing sample of $\{f_m\}_{m_1}^{m_R}$ and thus it is difficult to obtain an exact value of either $\overline{\mu}_{\tilde{A}}(f_m')$ or $\mu_{\tilde{U}}(f_m', \overline{\mu}_{\tilde{A}}(f_m'))$. To alleviate this problem, we consider

$$\overline{\mu}_{\tilde{A}}(f_m') = \overline{\mu}_{\tilde{A}}(f_{m_r}) \qquad (5.52.a)$$

and

$$\mu_{\tilde{U}}(f_m', \overline{\mu}_{\tilde{A}}(f_m')) = \mu_{\tilde{U}}(f_{m_r}, \overline{\mu}_{\tilde{A}}(f_{m_r})) \qquad (5.52.b)$$

where $f_{m_r} \in \{f_{m1}, f_{m2}, \ldots, f_{mR}\}$ is the closest feature point to $f_m'$.

Evaluation of (5.51) for $m = [1, M]$ and $c = [1, C]$ yields $C$, $M$-dimensional vectors, represented by

$$\overrightarrow{Q}^{\prime c} = [Q_1'^c, Q_2'^c, \ldots, Q_M'^c] \quad \text{for } c = [1, C] \qquad (5.53)$$

where each $\overrightarrow{Q}^{\prime c}$ symbolizes the membership vector of $F'$ to the respective pathological disorder class $K_c$.

After obtaining $\overrightarrow{Q}^{\prime c}$ corresponding to the unknown gestural expression $F'$, the extent of support of the pathological disorder class $K_c$ to $F'$ is determined by measuring

$$S^c = \frac{1}{\left\| \overrightarrow{Q}^c - \overrightarrow{Q}^{\prime c} \right\|} \qquad (5.54)$$

Here $\|.\|$ represents the Euclidean norm. The higher the value of $S^c$, higher is the certainty of $F'$ to fall in the respective pathological disorder category $K_c$. The appropriate pathological disorder class of $F'$ will be $K_{class}$, if $S^{class}$ is maximum over all $\{S^c\}_{c=1}^C$. The outline of identifying the pathological disorder class $K_{class}$ corresponding to the unknown gestural feature set $F' = \{f_1', f_2', \ldots, f_M'\}$. is given in Fig. 5.11 for $class \in [1, C]$.

**(a)**

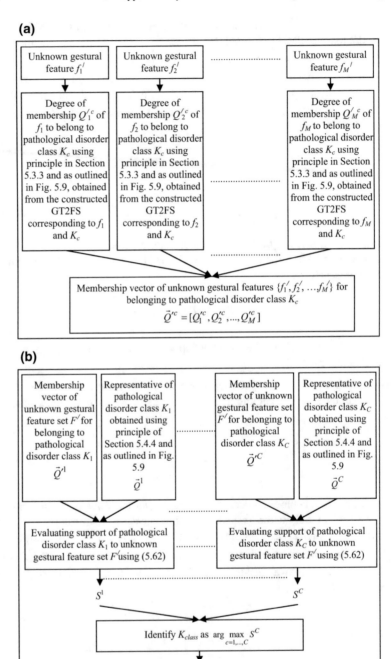

**Fig. 5.11 a** Evaluating membership vector of unknown expression to fall in class $K_c$. **b** Identifying class of unknown feature set $F'$

## 5.4 Experiment Details

Here, we present the experimental results of pathological disorder recognition using the principles introduced in Sect. 5.3. We consider $C$ (= 16) classes of pathological disorders as presented in Table 5.2. The experiment is conducted using the training data set **Tr**, used for designing the interval valued (or IT2) and GT2 fuzzy gesture space. It consists of $N$ (= 120) subjects, each exhibiting $L$ (= 600) gestural instances of each of the $C$ (= 16) pathological disorders. Hence, we have altogether $N \times L \times C = 120 \times 600 \times 16 = 1,152,000$ feature data points in **Tr**. For testing purpose, a separate dataset of 150 subjects each expressing gestures resembling different pathological disorders, is taken into account. $M$ (= 15) features are

**Table 5.2** Gesture space

| Disorder Name | Skeletal images obtained from the kinect sensor recorded twenty joint co-ordinates of the skeletons | | | |
| --- | --- | --- | --- | --- |
| | While sitting | | While standing | |
| | Pain at left joint | Pain at right joint | Pain at left joint | Pain at right joint |
| Osteoarthritis hand | | | | |
| Frozen shoulder | | | | |
| Plantar fasciitis | | | | |
| Tennis elbow | | | | |

extracted from each gestural expression to model the interval valued (or IT2) and GT2 fuzzy gesture space. This feature extraction process is carried out with the help of Microsoft's Kinect sensor (Xbox 360) and SDK v. 1.6.

### 5.4.1  Kinect Sensor

The basic principle of Microsoft's Kinect sensor [6, 38, 39] captures the twenty joint co-ordinates of the human skeletons at variable rate (for this chapter it is 30 fps) using RGB camera and Infra-Red (IR) depth sensor of a large viewing angle. The number of skeletal frames recorded for a specific gestural expression is $L = 600$ for 20 s duration. Thus the skeleton of a subject is represented by 20 body joint co-ordinates in 3-dimension space after combining the information from the RGB camera and the depth sensor under any ambient light conditions. The diagram of the Kinect sensor depicting all the components is given in Fig. 5.12.

### 5.4.2  Description of Pathological Disorders

The pathological disorders considered for this proposed system are the imitated gestural expressions shown by the subjects for different symptoms. In this chapter a wide range of disorders for elder persons are considered. The first concerned disorder is Cerviacal Spondylosis, which is a joint problem. The degenerative changes in the cervical spine cause pain usually occurring in elderly persons. Deterioration of joints in hand and knee causes Osteoarthritis hand and Osteoarthritis knee respectively, leading to pain and stiffness. The chapter also looks at the problems associated with muscle and tendon. Frozen shoulder is one type of disorder where the injury over the supporting muscle of the shoulder leads to stiffness of the shoulder for the elders. Strain in the tendons of ankle and elbow due to aging causes disorders like Plantar fasciitis and correspondingly. Hence the total number of specialized pathological disorder classes ($C$) considered in the chapter = number of generalized physical disabilities considered × number of postures × number of joints = 4 (Osteoarthritis hand, Osteoarthritis knee, Frozen shoulder and Tennis elbow) × 2 (sitting and standing) × 2 (left and right) = 16. It is elaborated in Table 5.2 (Fig. 5.14).

### 5.4.3  Feature Extraction

Total 15 features are extracted for the proposed algorithm for pathological disorder recognition using Kinect sensor. These extracted 15 features can be categorized into two classes—(a) the first 5 features are assembled under the group of distance

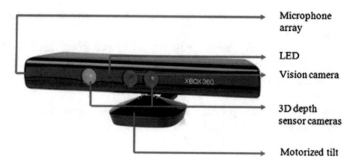

Microphone array

LED

Vision camera

3D depth sensor cameras

Motorized tilt

**Fig. 5.12**   Kinect sensor

features and (b) the remaining 10 belong to the group of angle features as shown in Fig. 5.13. The features are

1.  Distance between Hand Right and Hand Left ($D_{HRL}$)
2.  Distance between Elbow Right and Elbow Left ($D_{ERL}$)
3.  Distance between Knee Right and Knee Left ($D_{KRL}$)
4.  Distance between Ankle Right and Ankle Left ($D_{ARL}$)
5.  Distance between Spine and Head Left ($D_{SH}$)
6.  Angle between the lines joining Wrist Left to Elbow Left and Elbow Left to Shoulder Left respectively ($A_{WESL}$)
7.  Angle between the lines joining Wrist Right to Elbow Right and Elbow Right to Shoulder Rightrespectively ($A_{WESR}$)

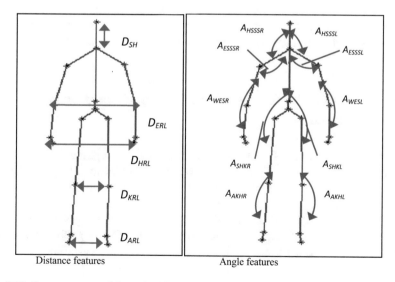

Distance features

Angle features

**Fig. 5.13**   Features extracted from the Kinect recorded skeleton

8. Angle between the lines joining Ankle Left to Knee Left and Knee Left to Hip Left respectively ($A_{AKHL}$)
9. Angle between the lines joining Ankle Right to Knee Right and Knee Right to Hip Rightrespectively ($A_{AKHR}$)
10. Angle between the lines joining Shoulder to Centre Spine and Hip Left to Knee Left respectively ($A_{SHKL}$)
11. Angle between the lines joining Shoulder to Centre Spine and Hip Right to Knee Rightrespectively ($A_{SHKR}$)
12. Angle between the lines joining Elbow Left to Shoulder Left and Shoulder Centre to Spine respectively ($A_{ESSSL}$)
13. Angle between the lines joining Elbow Right to Shoulder Right and Shoulder Centre to Spine respectively ($A_{ESSSR}$)
14. Angle between the lines joining Head to Shoulder Centre and Shoulder Centre to Shoulder Left respectively ($A_{HSSSL}$)
15. Angle between the lines joining Head to Shoulder Centre and Shoulder Centre to Shoulder Right respectively ($A_{HSSSR}$).

However, it is to be mentioned that the measurements of the features extracted from gestural expressions are mapped within the range [0, 10] to adopt the EIA methodology.

### 5.4.4   Dataset Construction

Experiments are undertaken on Kinect-recorded skeletons of three gender independent datasets of research fellows of Jadavpur University of ages in the range— (a) 25–30 years, (b) 30–35 years and (c) 35–40 years. Selective images from the gestural expressions of each dataset are given in Fig. 5.14. Training and test image data partitions for the experimental datasets are given in Table 5.3.

### 5.4.5   Creating the Interval Valued Fuzzy Gesture Space Using EIA

Initially, the interval valued fuzzy space, corresponding to a gestural feature $f_m$ and pathological disorder class $K_c$, contains $N = 120$ primary membership distributions obtained from the end point intervals $[a_n, b_n]$ (corresponding to $L$ measurements of $f_m$ of the $n$th subject in $\mathbf{Tr}^c$) for $n = [1, N]$. Due to pre-processing of EIA approach, $N^* < N$ end point intervals survive to construct the FOU for gestural feature $f_m$ corresponding to pathological disorder class $K_c$. The FOU of $f_m$ for pathological disorder class $K_c$ can be classified into one of the three categories, including

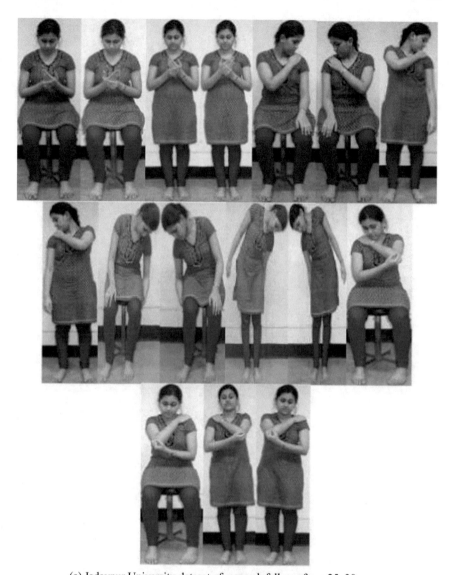

(a) Jadavpur University dataset of research fellow of age 25–30 years

**Fig. 5.14** Experiments done on the Jadavpur University dataset of research fellows

interior, left shoulder and right shoulder, based on the statistics given in (25) to (27). This is repeated for $m = [1, M]$ and $c = [1, C]$. Table 5.4 summarizes the number of end point intervals survived after each of the three pre-processing stages for each of the $M = 16$ gestural features corresponding to the training dataset **Tr** of

(b) Jadavpur University dataset of research fellow of age 30–35 years

**Fig. 5.14** (continued)

dataset 1 for pathological disorder: Tennis elbow at the left joint while standing. Table 5.4 also gives the final left and right end point statistics that are used to establish the nature of the FOU of each gestural feature of the same pathological disorder class, based on the data intervals obtained after all pre-processing stages. It

(c) Jadavpur University dataset of research fellow of age 35–40 years

**Fig. 5.14**  (continued)

is to be noted that the reported feature values in Table 5.4 are obtained by mapping the original measurements in the range [0, 10] to follow the principle of EIA.

For an example, we consider the case of feature $D_{HRL}$ for the same pathological disorder i.e., Tennis elbow at the left joint while standing. It is observed from

**Table 5.3**  Training and testing data set preparation

| Dataset | Training set (Tr) of gestural images $(N \times L \times C)$ | Testing gestural images |
|---|---|---|
| Research fellows of age 25–30 years | $120 \times 600 \times 16$ | 150 |
| Research fellows of age 30–35 years | $120 \times 600 \times 16$ | 150 |
| Research fellows of age 35–40 years | $120 \times 600 \times 16$ | 150 |

Table 5.4 that for $m_a = 1.977$, $m_b = 4.411$, $\sigma_c = 23.528$ and $\sigma_d = 4.974$. Hence $5.831 m_a - t_{\alpha, N^*-1} \frac{\sigma_c}{\sqrt{N^*}} = 4.083$ and $0.171 m_a + 8.29 - t_{\alpha, N^*-1} \frac{\sigma_d}{\sqrt{N^*}} = 7.058$ with $\alpha = 0.05$. Thus (26) is satisfied with this set of measurements of $D_{HRL}$ for pathological disorder: Tennis elbow at the left joint while standing. Consequently it yields a left shoulder FOU. Similar conclusion can be drawn from the illustrative plot in Fig. 5.15.

### 5.4.6   Creating the GT2 Fuzzy Gesture Space

After EIA pre-processing, the GT2 fuzzy space $\tilde{A}$, corresponding to a gestural feature $f_m$ and pathological disorder class $K_c$, is constructed. It contains 2 embedded T2FSs, $\tilde{U}$ and $\tilde{L}$, for UMF and LMF of $\tilde{A}$ respectively. Figure 5.16 shows the embedded T2FSs for gestural feature $D_{HRL}$ of dataset 1 and pathological disorder: Tennis elbow at the left joint while standing. This is done for $m = [1, M]$ and $c = [1, C]$.

### 5.4.7   Identifying Pathological Disorder Class Representatives

After obtaining the GT2 fuzzy gesture space $\tilde{A}$ for all $m = [1, M]$, the representatives of each pathological disorder class $K_c$ is identified using the feature data points of the training dataset **Tr** and following the principle as stated in Sects. 5.3.3 and 5.3.4. It is performed for all classes $c = [1, C]$. The $M = 16$-dimensional representatives for all $C = 16$ pathological disorder classes are listed in Table 5.5 corresponding to Jadavpur University dataset of research fellows of age 25–30 years only.

**Table 5.4** End point statistics and FOU category for sustaining intervals after EIA pre-processing for pathological disorder: tennis elbow at the left joint while standing using training dataset1 with $\alpha = 0.05$

| Features | Number of end point intervals after | | | Left end statistics | | Right end statistics | | $\sigma_c$ | $\sigma_d$ | FOU |
|---|---|---|---|---|---|---|---|---|---|---|
| | Outlier processing $N''$ | Tolerance limit processing $Y_2$ | Reasonable interval test $N^*$ | $m_a$ | $\sigma_a$ | $m_b$ | $\sigma_b$ | | | |
| $D_{HRL}$ | 75 | 58 | 41 | 1.977 | 3.945 | 4.411 | 4.919 | 23.52 | 4.974 | Left Shoulder |
| $D_{ERL}$ | 64 | 54 | 46 | 5.265 | 4.935 | 7.306 | 0.563 | 28.78 | 1.075 | Interior |
| $D_{KRL}$ | 73 | 64 | 43 | 5.034 | 1.293 | 6.883 | 2.838 | 8.061 | 2.849 | Interior |
| $D_{ARL}$ | 71 | 60 | 38 | 3.286 | 7.571 | 4.721 | 3.356 | 44.27 | 3.639 | Left Shoulder |
| $D_{SH}$ | 72 | 55 | 41 | 4.311 | 7.224 | 6.535 | 7.708 | 42.82 | 7.824 | Interior |
| $A_{WESL}$ | 67 | 62 | 49 | 5.097 | 5.577 | 8.062 | 6.156 | 33.09 | 6.243 | Interior |
| $A_{WESR}$ | 73 | 66 | 50 | 2.464 | 1.206 | 3.679 | 3.172 | 7.716 | 3.180 | Interior |
| $A_{AKHL}$ | 72 | 57 | 42 | 6.372 | 7.037 | 7.217 | 3.148 | 41.15 | 3.408 | Interior |
| $A_{AKHR}$ | 66 | 59 | 40 | 6.597 | 5.061 | 7.162 | 3.891 | 29.76 | 4.003 | Interior |
| $A_{SHKL}$ | 67 | 56 | 42 | 7.660 | 0.787 | 7.924 | 8.632 | 9.778 | 8.634 | Interior |
| $A_{SHKR}$ | 73 | 61 | 36 | 7.532 | 6.079 | 8.182 | 6.328 | 36.01 | 6.428 | Interior |
| $A_{ESSSL}$ | 75 | 61 | 48 | 4.892 | 7.600 | 6.599 | 8.742 | 45.17 | 8.855 | Interior |
| $A_{ESSSR}$ | 74 | 57 | 50 | 2.282 | 5.511 | 4.590 | 5.670 | 32.63 | 5.762 | Left Shoulder |
| $A_{HSSSL}$ | 68 | 63 | 44 | 5.791 | 0.959 | 7.726 | 4.936 | 7.461 | 4.940 | Interior |
| $A_{HSSSR}$ | 68 | 52 | 38 | 4.974 | 4.483 | 7.808 | 5.522 | 26.72 | 5.585 | Interior |

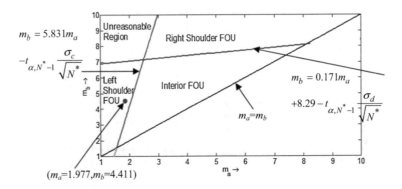

**Fig. 5.15** Graphical selection of FOU

**Fig. 5.16** GT2 fuzzy space
of gesture DHRL
corresponding to pathological
disorder: tennis elbow at the
*left* joint while standing using
training dataset 1

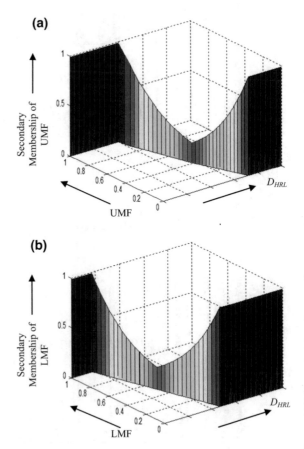

**Table 5.5** Class representative values

| Pathological disorders | | | Memberships of gestural features | | | | | | | | | | | | | | |
|---|---|---|---|---|---|---|---|---|---|---|---|---|---|---|---|---|---|
| Disorder | Sitting (St)/standing (Sd) | Joint | $D_{HRL}$ | $D_{ERL}$ | $D_{KRL}$ | $D_{ARL}$ | $D_{SH}$ | $A_{WESL}$ | $A_{WESR}$ | $A_{AKHL}$ | $A_{AKHR}$ | $A_{SHKL}$ | $A_{SHKR}$ | $A_{ESSL}$ | $A_{ESSR}$ | $A_{HSSSL}$ | $A_{HSSSR}$ |
| Osteoarthritis hand | St | Left | 0.83 | 0.73 | 0.93 | 0.87 | 0.98 | 0.88 | 0.90 | 0.99 | 0.87 | 0.74 | 0.89 | 0.94 | 0.98 | 0.97 | 0.91 |
| | St | Right | 0.86 | 0.93 | 0.75 | 0.97 | 0.81 | 0.89 | 0.86 | 0.94 | 0.88 | 0.93 | 0.79 | 0.88 | 0.87 | 0.92 | 0.94 |
| | Sd | Left | 0.73 | 0.89 | 0.95 | 0.87 | 0.74 | 0.94 | 0.93 | 0.88 | 0.99 | 0.91 | 0.97 | 0.82 | 0.83 | 0.74 | 0.77 |
| | Sd | Right | 0.77 | 0.90 | 0.74 | 0.86 | 0.79 | 0.92 | 0.88 | 0.99 | 0.90 | 0.91 | 0.94 | 0.93 | 0.77 | 0.78 | 0.82 |
| Frozen shoulder | St | Left | 0.86 | 0.84 | 0.88 | 0.86 | 0.77 | 0.79 | 0.86 | 0.85 | 0.87 | 0.99 | 0.87 | 0.75 | 0.87 | 0.89 | 0.86 |
| | St | Right | 0.78 | 0.83 | 0.89 | 0.78 | 0.84 | 0.88 | 0.75 | 0.86 | 0.89 | 0.85 | 0.93 | 0.91 | 0.72 | 0.90 | 0.75 |
| | Sd | Left | 0.96 | 0.73 | 0.96 | 0.93 | 0.98 | 0.93 | 0.79 | 0.94 | 0.95 | 0.77 | 0.99 | 0.80 | 0.85 | 0.93 | 0.87 |
| | Sd | Right | 0.98 | 0.88 | 0.86 | 0.84 | 0.95 | 0.89 | 0.83 | 0.82 | 0.80 | 0.77 | 0.74 | 0.88 | 0.82 | 0.95 | 0.93 |
| Plantar fasciitis | St | Left | 0.88 | 0.80 | 0.81 | 0.92 | 0.88 | 0.99 | 0.86 | 0.79 | 0.91 | 0.92 | 0.82 | 0.82 | 0.83 | 0.89 | 0.85 |
| | St | Right | 0.92 | 0.81 | 0.82 | 0.72 | 0.99 | 0.96 | 0.92 | 0.93 | 0.87 | 0.89 | 0.87 | 0.86 | 0.86 | 0.78 | 0.88 |
| | Sd | Left | 0.84 | 0.89 | 0.95 | 0.94 | 0.91 | 0.86 | 0.94 | 0.85 | 0.89 | 0.96 | 0.93 | 0.89 | 0.82 | 0.90 | 0.83 |
| | Sd | Right | 0.97 | 0.83 | 0.92 | 0.92 | 0.91 | 0.89 | 0.82 | 0.82 | 0.96 | 0.78 | 0.97 | 0.94 | 0.74 | 0.79 | 0.98 |
| Tennis elbow | St | Left | 0.96 | 0.82 | 0.79 | 0.82 | 0.90 | 0.80 | 0.91 | 0.85 | 0.86 | 0.90 | 0.90 | 0.92 | 0.91 | 0.84 | 0.95 |
| | St | Right | 0.83 | 0.94 | 0.92 | 0.72 | 0.76 | 0.91 | 0.93 | 0.91 | 0.87 | 0.89 | 0.88 | 0.93 | 0.77 | 0.84 | 0.97 |
| | Sd | Left | 0.91 | 0.83 | 0.97 | 0.81 | 0.80 | 0.85 | 0.89 | 0.90 | 0.98 | 0.84 | 0.83 | 0.95 | 0.87 | 0.96 | 0.87 |
| | Sd | Right | 0.99 | 0.94 | 0.96 | 0.81 | 0.82 | 0.82 | 0.93 | 0.93 | 0.86 | 0.84 | 0.90 | 0.80 | 0.80 | 0.84 | 0.98 |

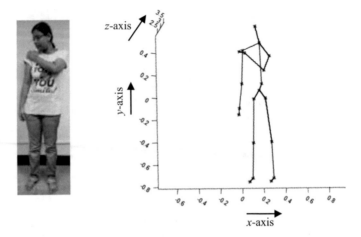

**Fig. 5.17**  Skeletal image of an unknown subject

## 5.4.8   *Pathological Disorder Recognition of an Unknown Gestural Expression*

The pattern recognition problem addressed here attempts to determine the pathological disorder of an unknown person from his/her gestural expression. The original feature values as well as their corresponding measurements after mapping into [0, 10], obtained from Fig. 5.17, are reported in Table 5.6. We now briefly explain the experimental results obtained by the reasoning methodology incorporating GT2 fuzzy gesture space of all pathological disorder classes. For sake of simplicity, the method is elaborated for Jadavpur University dataset of research fellows of age 25–30 years only.

The membership vectors of the unknown gestural feature set $F'$ (as reported in Table 5.6) to individual pathological disorder class are evaluated using the strategy proposed in Sect. 5.3.6 and are presented in Table 5.7. The support of each individual pathological disorder variant to the unknown feature set $F'$ is obtained by consulting Tables 5.5 and 5.7 and using (5.54). It is apparent from Table 5.8 that the maximum degree of support to $F'$ is given by the pathological disorder: Frozen shoulder at the right joint while standing. Hence $F'$ is assigned as a gestural feature member of this class.

**Table 5.6** Extracted set F′ of features of Fig. 5.17

Original measurements

| $D_{HRL}$ | $D_{ERL}$ | $D_{KRL}$ | $D_{ARL}$ | $D_{SH}$ | $A_{WESL}$ | $A_{WESR}$ | $A_{AKHL}$ | $A_{AKHR}$ | $A_{SHKL}$ | $A_{SHKR}$ | $A_{ESSSL}$ | $A_{ESSSR}$ | $A_{HSSSL}$ | $A_{HSSSR}$ |
|---|---|---|---|---|---|---|---|---|---|---|---|---|---|---|
| 0.55 | 0.27 | 0.15 | 0.17 | 0.55 | 58.5 | 146.4 | 172.6 | 173.2 | 166.4 | 168.3 | 12.5 | 14.1 | 144.2 | 126.0 |

Measurements mapped into [0, 10]

| $D_{HRL}$ | $D_{ERL}$ | $D_{KRL}$ | $D_{ARL}$ | $D_{SH}$ | $A_{WESL}$ | $A_{WESR}$ | $A_{AKHL}$ | $A_{AKHR}$ | $A_{SHKL}$ | $A_{SHKR}$ | $A_{ESSSL}$ | $A_{ESSSR}$ | $A_{HSSSL}$ | $A_{HSSSR}$ |
|---|---|---|---|---|---|---|---|---|---|---|---|---|---|---|
| 9.27 | 5.45 | 3.82 | 2.78 | 7.45 | 3.2 | 8.1 | 9.7 | 9.7 | 9.2 | 9.3 | 1.7 | 2.0 | 9.1 | 8.0 |

**Table 5.7** Membership vectors of the unknown gestural feature set $F'$

| Pathological disorders | | | | Memberships of gestural features | | | | | | | | | | | | | | |
|---|---|---|---|---|---|---|---|---|---|---|---|---|---|---|---|---|---|---|
| Disorder | St/Sd | Joint | $D_{HRL}$ | $D_{ERL}$ | $D_{KRL}$ | $D_{ARL}$ | $D_{SH}$ | $A_{WESL}$ | $A_{WESR}$ | $A_{AKHL}$ | $A_{AKHR}$ | $A_{SHKL}$ | $A_{SHKR}$ | $A_{ESSSL}$ | $A_{ESSSR}$ | $A_{HSSSL}$ | $A_{HSSSR}$ |
| Osteoarthritis hand | St | Left | 0.00 | 0.90 | 0.83 | 0.88 | 0.00 | 0.00 | 0.00 | 0.00 | 0.00 | 0.00 | 0.00 | 0.62 | 0.00 | 0.00 | 0.00 |
| | St | Right | 0.00 | 0.92 | 0.82 | 0.88 | 0.00 | 0.00 | 0.00 | 0.00 | 0.00 | 0.00 | 0.00 | 0.57 | 0.00 | 0.00 | 0.00 |
| | Sd | Left | 0.00 | 0.48 | 0.86 | 0.01 | 0.82 | 0.00 | 0.00 | 0.00 | 0.00 | 0.76 | 0.40 | 0.00 | 0.00 | 0.73 | 0.00 |
| | Sd | Right | 0.00 | 0.38 | 0.87 | 0.01 | 0.82 | 0.00 | 0.73 | 0.00 | 0.00 | 0.75 | 0.13 | 0.00 | 0.00 | 0.75 | 0.26 |
| Frozen shoulder | St | Left | 0.73 | 0.56 | 0.81 | 0.60 | 0.00 | 0.00 | 0.00 | 0.00 | 0.00 | 0.00 | 0.00 | 0.20 | 0.64 | 0.00 | 0.00 |
| | St | Right | 0.87 | 0.66 | 0.73 | 0.67 | 0.00 | 0.00 | 0.00 | 0.00 | 0.00 | 0.00 | 0.00 | 0.00 | 0.00 | 0.00 | 0.00 |
| | Sd | Left | 0.46 | 0.66 | 0.05 | 0.34 | 0.63 | 0.00 | 0.00 | 0.75 | 0.77 | 0.18 | 0.65 | 0.02 | 0.00 | 0.00 | 0.00 |
| | Sd | Right | 0.54 | 0.64 | 0.06 | 0.31 | 0.55 | 0.18 | 0.74 | 0.86 | 0.74 | 0.97 | 0.62 | 0.59 | 0.43 | 0.67 | 0.78 |
| Plantar fasciitis | St | Left | 0.00 | 0.28 | 0.82 | 0.43 | 0.00 | 0.00 | 0.00 | 0.00 | 0.00 | 0.00 | 0.00 | 0.00 | 0.29 | 0.00 | 0.00 |
| | St | Right | 0.00 | 0.31 | 0.92 | 0.46 | 0.00 | 0.00 | 0.00 | 0.00 | 0.00 | 0.00 | 0.00 | 0.00 | 0.00 | 0.39 | 0.00 |
| | Sd | Left | 0.47 | 0.00 | 0.70 | 0.54 | 0.00 | 0.02 | 0.00 | 0.00 | 0.00 | 0.00 | 0.00 | 0.80 | 0.00 | 0.00 | 0.00 |
| | Sd | Right | 0.36 | 0.00 | 0.68 | 0.69 | 0.00 | 0.00 | 0.00 | 0.20 | 0.29 | 0.00 | 0.00 | 0.46 | 0.00 | 0.00 | 0.00 |
| Tennis elbow | St | Left | 0.00 | 0.50 | 0.35 | 0.97 | 0.00 | 0.00 | 0.60 | 0.00 | 0.00 | 0.00 | 0.00 | 0.00 | 0.00 | 0.00 | 0.75 |
| | St | Right | 0.00 | 0.41 | 0.39 | 0.93 | 0.00 | 0.00 | 0.00 | 0.00 | 0.00 | 0.00 | 0.00 | 0.12 | 0.36 | 0.77 | 0.00 |
| | Sd | Left | 0.00 | 0.70 | 0.72 | 0.74 | 0.00 | 0.00 | 0.60 | 0.00 | 0.00 | 0.00 | 0.00 | 0.00 | 0.00 | 0.00 | 0.00 |
| | Sd | Right | 0.00 | 0.78 | 0.70 | 0.79 | 0.00 | 0.00 | 0.00 | 0.00 | 0.00 | 0.00 | 0.00 | 0.20 | 0.53 | 0.29 | 0.00 |

**Table 5.8** Supports of each class to the unknown gestural feature set $F'$

| Pathological disorder | | | Support |
|---|---|---|---|
| Disorder | Sitting/standing | Joint | S |
| Osteoarthritis hand | Sitting | Left | 0.3282 |
| | Sitting | Right | 0.3375 |
| | Standing | Left | 0.3689 |
| | Standing | Right | 0.3846 |
| Frozen shoulder | Sitting | Left | 0.3701 |
| | Sitting | Right | 0.3524 |
| | Standing | Left | 0.3908 |
| | Standing | Right | 0.6680 |
| Plantar fasciitis | Sitting | Left | 0.3273 |
| | Sitting | Right | 0.3225 |
| | Standing | Left | 0.3306 |
| | Standing | Right | 0.3474 |
| Tennis elbow | Sitting | Left | 0.3452 |
| | Sitting | Right | 0.3415 |
| | Standing | Left | 0.3337 |
| | Standing | Right | 0.3495 |

## 5.5   Performance Analysis

The performance of the proposed T2 fuzzy gesture space induced pathological disorder recognition algorithm is examined here with respect to three datasets considered in this chapter.

### 5.5.1   Comparative Framework

This chapter compares the relative performance of the proposed T2FS algorithms with 7 traditional gesture recognition algorithms/techniques. The comparative framework includes traditional IVFS [16, 18], traditional T1FS, support vector machine (SVM) classifier [4, 5], $k$-nearest neighbour ($k$-NN) classification [7, 11], Levenberg–Marquardt algorithm induced neural network (LMA-NN) [40], ensemble decision tree (EDT) [41], and radial basis function network (RBFN) [28, 42].

For traditional IVFS based gesture recognition scheme, an interval valued fuzzy gesture space is created for each gestural feature $f_m$ and pathological disorder class $K_c$ following the principle given in Sect. 5.3.1. Hence we obtain $M$ IVFSs under each of the $C$ pathological disorder class head. The centroid of each IVFS is evaluated by taking average of its respective left and right centroids. Thus each pathological disorder class is represented by its $M$-dimensional centroid vector. The unknown gesture (encoded by a set of $M$ extracted features) is assigned to the pathological disorder class providing minimal discrepancy between its respective centroid vector representative and the unknown gesture.

In the traditional T1FS based approach, a T1 fuzzy gesture space is created for each gestural feature $f_m$ and pathological disorder class $K_c$. A symmetric triangular T1FS corresponding to $f_m$ and $K_c$ is constructed by considering minimum and maximum values of $f_m$ of all feature data belonging to class $K_c$ in the training dataset. Then as in previous case, here also, the T1 defuzzified centroid of the T1FS, corresponding to each $f_m$ and $K_c$, is evaluated. It assists in representing a class by an $M$-dimensional centroid. The unknown gesture now can be identified by the same method as in traditional IVFS.

In case of SVM (RBF kernel) based approach, the classifier is trained to partition each pair of $M$-dimensional feature points, belonging to two different classes, by maximizing the distance between the support vectors of two classes. The class yielding highest value of RBF kernel is declared as the owner of the unknown gesture.

The $k$-nearest neighbours ($k$-NN) algorithm selects the $k$ feature vectors (each assigned with its respective class label $c \in [1, C]$) from the training dataset which are nearest to the unknown gestural expression. The nearest neighbours are identified using Euclidean distance. Then following the majority voting principle, the pathological disorder class label is identified whose members are most frequently present in the list of $k$ nearest neighbours. It is then declared as the class label of the unknown gesture. Here $k$ is taken as 10. The performance of the back-propagation assisted artificial neural network is also studied in this context. Here the neural network weights are adapted using Levenberg-Marquardt Optimization. This technique incorporates a mixture of the simple gradient descent learning and quadratic approximations that helps to achieve small error at a faster pace.

An ensemble classifier or a multiple classifier system is made up of a number of 'base' classifiers or 'weak learners'. Each of the constituent classifier classifies the training data set separately. The outcome of the ensemble classifier is a combination of the decisions of these base classifiers. A 'tree' classifier has been used as the weak learner in this work. Each node of the tree classifier decides on each of the features in the dataset to predict the class of a sample. Two popular methods to implement the ensemble classifiers are bagging and boosting. The present work has utilized adaptive boosting algorithm (AdaBoost) to implement a boosting ensemble classifier based on the tree learning. Initially the weights of all the samples in the dataset are equal. In each iteration, a new weak classifier is created using the whole dataset. Moreover, the weights for all samples are increased (or decreased) based on the samples misclassified (or correctly classified). Hence it is of 'adaptive' nature. In our comparative framework, another popular artificial neural network, RBFN, is considered. The activation function of RBFN basically resembles a radial basis function.

## 5.5.2 Parameter Settings

For all the contestant algorithms, we have used a common framework in terms of their features and datasets. We employ the best parametric set-up for all these algorithms as prescribed in their respective sources. The effectiveness of the proposed T2FS based gesture recognition greatly depends on the value of scaling parameter $\alpha$ of (37). Again, $\alpha$ is determined from the pre-specified minimum secondary membership $\mu_{\underset{U}{\sim}}^{min}$ of (37). After several trials we have noted that our proposed T2FS based gesture recognition strategy attains its highest accuracy for $\mu_{\underset{U}{\sim}}^{min} = 0.3$ and hence $\mu_{\underset{U}{\sim}}^{min}$ is kept at 0.3.

## 5.5.3 Performance Metric

The pathological disorder $c$ of a subject (patient) in reality and that recognized by an algorithm may have four combinations as enlisted below.

*True Positive of class c (TP_c):*     the number of pathological disorders that are recognized as in class $c$ and indeed belong to class $c$

*False Positive of class c (FP_c):*     the number of pathological disorders that are recognized as in class $c$ but in fact belong to class $c' \in [1, C]$ provided $c' \neq c$

*False Negative of class c (FN_c):*     the number of pathological disorders that are recognized as in class $c' \in [1, C]$ provided $c' \neq c$ but indeed belong to class $c$

*True Negative of class c (TN_c):*     the number of pathological disorders that are correctly recognized to not to belong to class $c$.

In order to allow a quantitative assessment in performance of different gesture mediated pathological disorder recognition algorithms, the comparative study uses the following performance metricsto evaluate the performance of a multi-class classification algorithm with $C$ classes (here $C = 16$) using the above four states and considering the micro-averaging strategy given in [38].

(a) **Recall**: It measures the effectiveness of a classifier to identify the class labels. Hence it relates to the tests ability to identify positive results.

$$Recall = \frac{\sum_{c=1}^{C} TP_c}{\sum_{c=1}^{C} (TP_c + FN_c)} \tag{5.55}$$

(b) **Precision**: It measures the degree of agreement of the data class labels with those of a classifier. A high precision for a given test means that when the test yields a positive result, it is most likely correct in its assessment.

$$Precision = \frac{\sum_{c=1}^{C} TP_c}{\sum_{c=1}^{C} (TP_c + FP_c)} \tag{5.56}$$

(c) **Accuracy**: It is the overall correctness of the predictive model and is calculated as the sum of correct *predictions* divided by the total number of predictions.

$$Accuracy = \sum_{c=1}^{C} \frac{TP_c + TN_c}{TP_c + TN_c + FP_c + FN_c} \tag{5.57}$$

(d) **F1_score**: It is interpreted as a weighted average of the precision and recall, both of which are related to the performance of the positive instances. An $F_1$ score reaches its best value at 1 and worst score at 0.

$$F1\_score = \frac{2 \times Precision \times Recall}{Precision + Recall} \tag{5.58}$$

(e) **Average Error Rate**: It is, in essence, the average per-class classification error.

$$Average\ Error\ Rate = \sum_{c=1}^{C} \frac{FP_c + FN_c}{TP_c + TN_c + FP_c + FN_c} \tag{5.59}$$

## 5.5.4  *Results and Performance Analysis*

The mean and the standard deviation (within parenthesis) of the best-of-run values of the performance metrics for 25 independent runs of each of the algorithm are presented in Table 5.9. We also use paired $t$ test to compare the means of each performance metric produced by the best and the second best algorithm. In the last column of Table 5.9 we represent the statistical significance level of the difference of the means of the best two algorithms. Note that here '+' indicates that the $t$ value of 49 degrees of freedom is significant at a 0.05 level of significance by two-tailed test, whereas '−' means the difference of means is not statistically significant, and 'NA' stands for not applicable, covering cases for which two or more algorithms achieve the best accuracy results. For all the t-tests carried in Table 5.9, the sample size is taken to be 25. The best algorithm is marked in bold.

**Table 5.9** Comparison of different recognition algorithms for 25 runs

| Dataset | Performance metrics | Algorithms | | | | | | | | Statistical significance |
|---|---|---|---|---|---|---|---|---|---|---|
| | | T2FS | Traditional IVFS | Traditional T1FS | SVM | k-NN | LMA-NN | EDT | RBFN | |
| Research fellows of age 25–30 years | Recall | 0.82 (0.1) | 0.79 (0.2) | 0.78 (0.4) | 0.74 (0.4) | 0.72 (0.5) | 0.70 (0.6) | 0.68 (0.6) | 0.63 (0.7) | + |
| | Precision | 0.91 (0.1) | 0.80 (0.1) | 0.80 (0.1) | 0.80 (0.1) | 0.79 (0.2) | 0.77 (0.5) | 0.60 (0.8) | 0.59 (0.9) | + |
| | Accuracy | 0.83 (0.1) | 0.82 (0.1) | 0.82 (0.2) | 0.81 (0.2) | 0.79 (0.2) | 0.77 (0.3) | 0.70 (0.5) | 0.70 (0.6) | + |
| | F1_score | 0.86 (0.1) | 0.84 (0.1) | 0.77 (0.2) | 0.77 (0.3) | 0.76 (0.5) | 0.73 (0.5) | 0.64 (0.8) | 0.61 (0.8) | + |
| | Average error Rate | 0.17 (0.1) | 0.17 (0.1) | 0.17 (0.1) | 0.18 (0.1) | 0.20 (0.4) | 0.22 (0.6) | 0.29 (0.7) | 0.29 (0.8) | NA |
| Research fellows of age 30–35 years | Recall | 0.84 (0.1) | 0.75 (0.2) | 0.76 (0.2) | 0.77 (0.2) | 0.73 (0.2) | 0.69 (0.3) | 0.73 (0.6) | 0.71 (0.8) | + |
| | Precision | 0.87 (0.1) | 0.84 (0.1) | 0.80 (0.1) | 0.84 (0.1) | 0.79 (0.1) | 0.77 (0.3) | 0.60 (0.5) | 0.58 (0.6) | + |
| | Accuracy | 0.84 (0.2) | 0.83 (0.2) | 0.82 (0.2) | 0.83 (0.2) | 0.80 (0.3) | 0.78 (0.5) | 0.71 (0.7) | 0.71 (0.7) | + |
| | F1_score | 0.86 (0.1) | 0.79 (0.1) | 0.78 (0.1) | 0.78 (0.2) | 0.76 (0.2) | 0.73 (0.4) | 0.66 (0.5) | 0.33 (0.8) | + |
| | Average error rate | 0.17 (0.1) | 0.16 (0.1) | 0.17 (0.1) | 0.17 (0.1) | 0.19 (0.3) | 0.21 (0.4) | 0.28 (0.8) | 0.28 (0.8) | – |

(continued)

**Table 5.9** (continued)

| Dataset | Performance metrics | Algorithms | | | | | | | | | Statistical significance |
|---|---|---|---|---|---|---|---|---|---|---|---|
| | | T2FS | Traditional IVFS | Traditional T1FS | SVM | k-NN | LMA -NN | EDT | RBFN | |
| Research fellows of age 35–40 years | Recall | 0.90 (0.1) | 0.79 (0.2) | 0.77 (0.2) | 0.77 (0.2) | 0.73 (0.3) | 0.73 (0.4) | 0.69 (0.8) | 0.64 (0.8) | + |
| | Precision | 0.93 (0.1) | 0.84 (0.2) | 0.81 (0.2) | 0.84 (0.2) | 0.8 (0.25) | 0.78 (0.3) | 0.60 (0.5) | 0.63 (0.6) | + |
| | Accuracy | 0.93 (0.1) | 0.86 (0.1) | 0.84 (0.1) | 0.81 (0.2) | 0.81 (0.4) | 0.78 (0.4) | 0.72 (0.6) | 0.72 (0.7) | + |
| | F1_score | 0.91 (0.2) | 0.82 (0.2) | 0.80 (0.3) | 0.79 (0.4) | 0.76 (0.4) | 0.76 (0.5) | 0.64 (0.5) | 0.64 (0.8) | + |
| | Average error rate | 0.12 (0.1) | 0.13 (0.3) | 0.14 (0.3) | 0.18 (0.3) | 0.18 (0.5) | 0.21 (0.6) | 0.27 (0.8) | 0.27 (0.8) | + |

A close inspection of Table 5.9 indicates that the performance of the proposed T2FS-based pathological disorder recognition algorithm has remained consistently superior to that of the other competitor methods. It is interesting to see that out of 5 performance metrics for all 3 datasets, i.e., out of 15 instances, in 13 cases, T2FS outperforms its nearest neighbour competitor in a statistically significant manner. Here traditional IVFS-based prediction method remains the second best algorithm.

Four statistical tests are undertaken to analyse the relative performance of the proposed algorithm over the existing ones.

### 5.5.5   McNemar's Statistical Test

Let algorithms $A$ and $B$ have a common training set **Tr** of gestures of pathological disorders.

Let,

$n_{01}$   be the number of pathological disorders misclassified by $A$ but not by $B$,
$n_{10}$   be the number of pathological disorders misclassified by $B$ but not by $A$,
$p_1$   be the marginal probability of the outcomes of the experiment where the examples are misclassified by algorithm $A$ but not by $B$, and
$p_2$   be the marginal probability of the outcomes of the experiment where the examples are misclassified by algorithm $B$ but not by $A$

The null hypothesis $H_0$ of the McNemar's test states that these two probabilities are equal. Symbolically, the null and the alternative hypotheses $H_0$ and $H_1$ are respectively represented as follows $H_0$: $p_1 = p_2$ and $H_1$: $p_1 \neq p_2$.

Then, under the null hypothesis that both algorithms have the same error rate, the McNemar's statistic $Z$, defined in (60), follows a $\chi^2$ with degree of freedom equals to 1 [43].

$$Z = \frac{(|n_{01} - n_{10}| - 1)^2}{n_{01} + n_{10}} \tag{5.60}$$

Let $A$ be the proposed T2FS algorithm and $B$ is one of the other seven contender algorithms. We thus evaluate $Z = Z_1$ through $Z_7$, where $Z_i$ denotes the comparator statistic of misclassification between the T2FS (Algorithm: $A$) and the $i$th of the seven algorithms (Algorithm: $B$), where the suffix $i$ refers to the algorithm in the $i$th row number $i$ of Table 5.10. In Table 5.10, the null hypothesis $H_0$ has been rejected, if $Z_i > \chi^2_{1, \alpha=0.05} = 3.84$, where $\chi^2_{1, \alpha=0.05} = 3.84$ is the critical value of the $\chi^2$ distribution for 1 degree of freedom at probability of $\alpha = 0.05$ [44].

It is apparent from Table 5.10 that the value of the McNemar's statistic $Z$, corresponding to the most of the contender algorithms (except for the traditional IVFS for Jadavpur University dataset of research scholars of age 25–30) and the proposed T2FS algorithm (based pathological disorder recognition) is greater than the critical value of 3.84. It indicates a statistically significant difference in the

**Table 5.10** Performance analysis using McNemar's test

| Competitor algorithm = B | Control algorithm A = T2FS | | | | | | | | | | | |
| --- | --- | --- | --- | --- | --- | --- | --- | --- | --- | --- | --- | --- |
| | Jadavpur university dataset of research scholars of age 25–30 | | | | Jadavpur University dataset of research scholars of age 30–35 | | | | Jadavpur university dataset of research scholars of age 35–40 | | | |
| | $n_{01}$ | $n_{10}$ | $Z_i$ | Comment on acceptance/rejection of null hypothesis | $n_{01}$ | $n_{10}$ | $Z_i$ | Comment on acceptance/rejection of null hypothesis | $n_{01}$ | $n_{10}$ | $Z_i$ | Comment on acceptance/rejection of null hypothesis |
| Traditional IVFS | 45 | 65 | 3.28 | Accept | 45 | 67 | 3.93 | Reject | 43 | 69 | 5.58 | Reject |
| Traditional T1FS | 45 | 68 | 4.28 | Reject | 45 | 68 | 4.28 | Reject | 39 | 69 | 7.78 | Reject |
| SVM | 37 | 68 | 8.57 | Reject | 43 | 68 | 5.18 | Reject | 36 | 73 | 11.88 | Reject |
| kNN | 33 | 72 | 13.75 | Reject | 32 | 76 | 17.12 | Reject | 35 | 77 | 15.00 | Reject |
| LMA-NN | 33 | 81 | 19.37 | Reject | 27 | 84 | 28.25 | Reject | 33 | 76 | 16.18 | Reject |
| EDT | 27 | 83 | 27.50 | Reject | 24 | 83 | 31.43 | Reject | 29 | 83 | 25.08 | Reject |
| RBFN | 24 | 82 | 30.65 | Reject | 25 | 85 | 31.64 | Reject | 26 | 82 | 28.00 | Reject |

performance between each competitor algorithm and our proposed T2FS based method for gesture recognition for $\alpha = 0.05$. Hence, it is apparent from Tables 5.9 and 5.10 that our proposed method outperforms the existing techniques in a statistically significant fashion according to the McNemar's test.

### 5.5.6 Friedman Test

A non-parametrical statistical test, known as the Friedman two-way analysis of variances by ranks [45], is also performed on the mean of *Accuracy* metric values for 25 independent runs of each of the eight algorithms as reported in Table 5.9 for all three datasets. The null hypothesis for the Friedman test states that there are no differences between the medians of the *Accuracy* measurements of three datasets obtained by each contender algorithm. Let $k$ be the number of competitor algorithms and $M_i$ be the median value of *Accuracy* obtained by the $i$th algorithm over three datasets for $i = [1, k]$. Symbolically, the null and the alternative hypotheses $H_0$ and $H_1$ are respectively represented as follows.

$H_0$: $M_1 = M_2 = \ldots = M_k$ indicating there is no significant difference between the performance of all $k$ algorithms.
$H_1$: Not all $M_i$s are equal for $i = [1, k]$ indicating there is a significant difference between the performance of the algorithms.

The Friedman test is carried out in the present context by the following procedure. Let $r_i^j$ be the ranking of the observed *Accuracy* obtained by the $i$th algorithm for the $j$th dataset. The best of all the $k$ algorithms is assigned a rank of 1 and the worst is assigned the ranking $k$. Table 5.11 summarizes the rankings obtained by the Friedman procedure. The average ranking of each individual algorithm is then obtained by averaging their individual ranks over the $N = 3$ datasets of Jadavpur University research scholars. Symbolically, the average ranking acquired by the $i$th algorithm over all $j = [1, N]$ datasets is defined as follows.

$$R_i = \frac{1}{N} \sum_{j=1}^{N} r_i^j \tag{5.61}$$

Under the null hypothesis, formed from the supposition that the results of the algorithms are equivalent and, therefore, their rankings are also similar, the Friedman's statistic, defined in (5.62), follows a $\chi_F^2$ distribution with degree of freedom equals to $k - 1$ [45].

$$\chi_F^2 = \frac{12N}{k(k+1)} \left[ \sum_{i=1}^{k} R_i^2 - \frac{k(k+1)^2}{4} \right] \tag{5.62}$$

**Table 5.11** Performance analysis using Friedman and Iman-Davenport tests

| Algorithm | Ranking $r_i^j$ of the $i$th algorithm obtained from the $j$th Jadavpur University dataset of research scholars of age | | | Average ranking $R_i$ |
|---|---|---|---|---|
| | $25$–$30$ $(r_i^1)$ | $30$–$35$ $(r_i^2)$ | $35$–$40$ $(r_i^3)$ | |
| T2FS | 1 | 1 | 1 | 1.00 |
| Traditional IVFS | 2 | 2 | 2 | 2.00 |
| Traditional T1FS | 3 | 4 | 3 | 3.33 |
| SVM | 4 | 3 | 4 | 3.67 |
| kNN | 5 | 5 | 5 | 5.00 |
| LMA-NN | 6 | 6 | 6 | 6.00 |
| EDT | 7 | 7 | 7 | 7.00 |
| RBFN | 8 | 8 | 8 | 8.00 |

| Friedman Test | | Iman-Davenport Test | |
|---|---|---|---|
| Statistics $\chi_F^2$ | Comment on acceptance/rejection of null hypothesis | Statistics $F_F$ | Comment on acceptance/rejection of null hypothesis |
| 20.7779 | Reject | 187.0946 | Reject |
| Critical difference for $\alpha = 0.05$ | | 2.606 | |

In this chapter, $N$ = number of datasets considered = 3 and $k$ = number of competitor algorithms = 8. In Table 5.11, it is shown that the null hypothesis has been rejected, as $\chi_F^2 = 20.7779$ is greater than $\chi_{7,\alpha=0.05}^2 = 14.067$, where $\chi_{7,\alpha=0.05}^2 = 14.067$ is the critical value of the $\chi_F^2$ distribution for $k - 1 = 7$ degrees of freedom at probability of $\alpha = 0.05$ [46].

It is evident from Table 5.11 that the proposed T2FS based pathological disorder recognition scheme has attained the best average Friedman rank among all the contenders for all three variants of datasets. It in turn indicates that the proposed T2FS algorithm outperforms other existing methods of gesture recognition irrespective of the dataset used.

### 5.5.7  Iman-Davenport Statistical Test

Additionally, we use Iman-Davenport test as a variant of Friedman test that provides better statistics [45]. It is a deviation from the Friedman's statistic given that the metric in (63) produces a conservative undesirably effect. The null and the alternative hypothesis of the Iman-Davenport test are given by

$H_0$: The accuracies in recognizing the pathological disorders by each of the $k$ contender algorithms are equivalent.

$H_1$: There exists significant difference between the accuracies in recognizing the pathological disorders by the $k$ algorithms.

The statistic proposed by Iman-Davenport is given as follows.

$$F_F = \frac{(N-1) \times \chi_F^2}{N \times (k-1) - \chi_F^2} \tag{5.63}$$

It is distributed according to $F_F$ distribution with $(k-1)$ and $(k-1) \times (N-1)$ degrees of freedom. For $k = 8$ and $N = 3$, it is shown in Table 5.11 that the null hypothesis has been rejected, as $F_F = 187.0946$ is greater than $F_{7,14,\alpha=0.05} = 3.38$, where $F_{7,14,\alpha=0.05} = 3.38$ is the critical value of the $F$ distribution for $k - 1 = 7$ and $(k-1) \times (N-1) = 14$ degrees of freedom at probability of $\alpha = 0.05$ [46].

It is apparent from Table 5.11 that both the Friedman and the Iman-Davenport statistics of 20.7779 and 187.0946 are greater than their respective critical values of 14.067 and 3.38. Accordingly, the null hypothesis concerned with the equivalent performance of all competitive algorithms is rejected. In other words, the significant superiority of the proposed T2FS algorithm over other competitors is substantiated from its Friedman rank of 1 and also from the statistic values of the Friedman and the Iman-davenport tests.

### 5.5.8 Bonferroni-Dunn Test

The results in Table 5.11 highlight T2FS as the best algorithm, so the post hoc analysis [47] is performed with T2FS as the control method. In the post hoc analysis we apply the Bonferroni-Dunn test [47] over the results of Friedman procedure. The null and the alternative hypothesis for the Bonferroni-Dunn test are given by

$H_0$: There exist no statistically significant difference between the performances of the control algorithm and any member of the remaining competitive algorithms.
$H_1$: There exists a statistically significant difference between the control algorithm and any member of the remaining competitive algorithms.

The analysis indicates the level of significance at which the control algorithm is better than each of the remaining algorithms (i.e., when the null hypothesis is rejected). For the Bonferroni-Dunn test, a critical difference $(CD)$ [46] is calculated which for these data comes as 2.606. The interpretation of this measure is that the performance of two algorithms is significantly different, only if their corresponding average Friedman ranks differ by at least a critical difference, which is pictorially depicted in Fig. 5.18. In Fig. 5.18, the height of each bar corresponds to the average Friedman ranking obtained for a contender algorithm (based on the accuracy of the pathological disorder recognition). Evidently, the smallest bar represents the best algorithm (here T2FS).

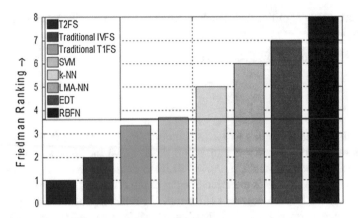

**Fig. 5.18** Graphical representation of Bonferroni-Dunn's procedure

To identify whether the difference between the average Friedman rank of the control algorithm (T2FS) and any one of the remaining seven contender algorithms exceeds $CD = 2.606$, a cut line is drawn (shown as a bold blue line) going through the bars. The height of the cut line = the height of the smallest bar (i.e., the average Friedman rank of the control algorithm T2FS) + $CD = 1 + 2.606 = 3.606$. It is evident that a bar (of height <3.606) lying below the cut line represents that the average Friedman rank of the corresponding algorithm differs from that of the control algorithm by less than 2.606 and hence, the performances of these two algorithms are considered to be statistically equivalent. Alternatively, we can conclude that the behaviours of the contender algorithms, characterized by the bars (of height >3.606) above the cut line, are significantly worse than the contributed by the control algorithm.

It can be perceived that only for traditional IVFS (Bonferroni-Dunn statistic = $|2 - 1| = 1 < CD$) and T1FS (Bonferroni-Dunn statistic = $|3.333 - 1| = 2.333 < CD$), the null hypothesis cannot be rejected with any of the tests for the level of significance $\alpha = 0.05$. It implies an insignificant statistical difference in the performance between the proposed T2FS, the traditional IVFS and T1FS based pathological disorder recognition policies. The other five algorithms (SVM, $k$NN, LMA-NN, EDT and RBFN), however, are regarded as significantly poorer than the T2FS (with respective Bonferroni-Dunn statistics of 2.667, 4, 5, 6 and 7, each greater than $CD$).

### 5.5.9  Robustness Analysis

This experiment undertakes robustness analysis of the proposed T2FS approach of gesture induced pathological disorder recognition with respect to the number of feature data points $N$ (initially) of training dataset **Tr**. Figure 5.19 shows the evolution of accuracy *acc*, computational complexity *cc* (referring to the total time required in seconds for the entire training phase) and the ratio $v = acc/cc$ (of the proposed method for gesture recognition) with $N$ for all three datasets of Jadavpur University research scholars. It is evident from the figure that the accuracy increases as the feature data points in the training datasets increase at the cost of computational overhead. However, a close observation of Fig. 5.19 reveals that $v$ is the highest for $N = 120$, irrespective of datasets. Motivated by this observation the optimal value of $N$ is set at 120.

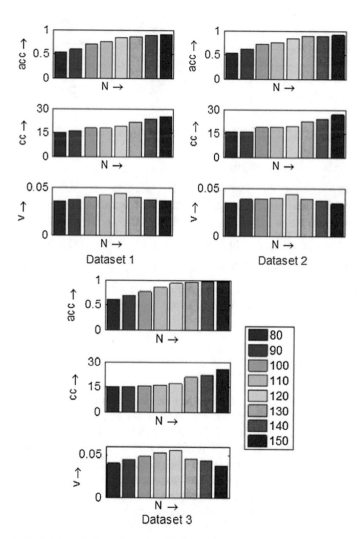

**Fig. 5.19** Robustness analysis of the proposed algorithm

**Table 5.12** Variation in performance of the proposed T2FS algorithm with the minimum setting of secondary membership grade

| Minimum secondary membership value $\mu_{\underset{U}{\sim}}^{min}$ | $Accuracy = \sum_{c=1}^{C} \frac{TP_c + TN_c}{TP_c + TN_c + FP_c + FN_c}$ (for $C = 16$ pathological disorder classes) obtained for Jadavpur University dataset of research scholars of age | | |
|---|---|---|---|
| | 25–30 | 30–35 | 35–40 |
| 0.10 | 0.7574 | 0.7246 | 0.8326 |
| 0.15 | 0.7600 | 0.7552 | 0.8612 |
| 0.20 | 0.7868 | 0.7929 | 0.9103 |
| 0.25 | 0.8020 | 0.8103 | 0.9223 |
| 0.30 | 0.8395 | 0.8455 | 0.9341 |
| 0.35 | 0.8303 | 0.8427 | 0.9128 |
| 0.40 | 0.8218 | 0.8239 | 0.9027 |

## 5.5.10  Effect of Minimum Secondary Membership Function Value

The performance of the proposed T2FS algorithm for pathological disorder recognition is based on the accurate selection of the minimum secondary membership $\mu_{\underset{U}{\sim}}^{min}$ which determines the parameter $\alpha$ in (5.37). We report our observation in terms of the *Accuracy* by T2FS method in Table 5.12 by selectively tuning $\mu_{\underset{U}{\sim}}^{min}$ from 0.1 to 0.4 for all three datasets. Experimental results indicate that when $\mu_{\underset{U}{\sim}}^{min}$ is set to 0.3, the performance of the algorithm is boosted up (marked in boldface). Hence it is kept at 0.3 for all three datasets.

## 5.6  Conclusion

The chapter presents a novel approach pathological disorder recognition system for elderly healthcare particularly in homes with fewer or no inmates by utilizing the composite benefit of GT2FS and IVFS (or IT2FS). The present work has aimed at gesture driven detection of the pathological disorder at early stage at home, rather than at hospital, to reduce the labor as well as the economical burden on our society. In order to classify an unknown gestural expression, the system makes use of the background knowledge about a large gesture dataset with known pathological disorder classes. The proposed work is concerned with identifying $C = 16$ pathological disorders from $M = 15$ features, extracted from the gestural expressions of patients. To accomplish this, the following steps are implemented.

1. A training dataset is created by measuring $L = 600$ instances of a specific gestural feature $f_m$ (using Kinect sensor), $m \in [1, M]$, belonging to a specific pathological disorder class $K_c$, $c \in [1, C]$, of a particular subject, say $n$. This is then repeated for all $N$ subjects, i.e., $n = [1, N]$, all suffering from the same pathological disorder $K_c$. This results in $N$ intervals (each interval for $L$ instances of feature $f_m$ of each subject) of the measured gestural feature $f_m$ of class $K_c$.

2. Next using the EIA approach, an interval valued fuzzy gesture space is created, comprising $N^* < N$ intervals (due to outlier processing, tolerance limit processing and reasonable interval test) of feature $f_m$ and class $K_c$. The FOU corresponding to feature $f_m$ and class $K_c$ can fall under one of the three categories, including interior, left shoulder, and right shoulder.

3. After that a GT2 fuzzy gesture space is created by modeling the secondary membership of a given gestural feature $f_m$ based on the primary membership values of the UMF and LMF of the FOU corresponding to the pathological disorder class $K_c$. It helps in the assessment of the certainty of assigning primary memberships to the given UMF and LMF. Two policies are adopted for secondary membership evaluation of the UMF and the LMF of a FOU. First, the maximum and minimum primary membership values of UMF and LMF are assigned with the highest secondary membership grade of unity. Second, the secondary membership at other measurements of feature $f_m$ exponentially falls off with its distance away from the feature point corresponding to the optimal primary membership value of UMF and LMF respectively.

4. Next a novel approach is used to identify the degree of primary membership of a measured gestural feature $f_m$ to belong to the pathological disorder class $K_c$. This is obtained by an aggregated measure of the degree of support (i.e., the secondary memberships) of the UMF and the LMF to the measured feature $f_m$.

5. The entire process of step (1) to (4) is repeated for all features $f_m$ and pathological disorder classes $K_c$, i.e., $m = [1, M]$ and $c = [1, C]$. It results in $C$, $M$-dimensional class representative vectors. The $m$th component of the representative vector of class $K_c$ denotes the degree of primary membership of feature $f_m$ to class $K_c$ for $m = [1, M]$ and $c = [1, C]$.

6. To identify an unknown pathological disorder of a patient, first $M$ features are extracted from his/her gestural expression. Then the degree of primary membership of the extracted feature $f_m'$ to belong to physical disorder class $K_c$ is identified using step (4) and consulting the GT2 fuzzy gesture space of feature $f_m$ and class $K_c$, for $m = [1, M]$ and $c \in [1, C]$. This results in the $M$-dimensional membership vector of the extracted feature set $\{f_1', f_2', \ldots, f_M'\}$ of the unknown gestural expression to the class $K_c$.

7. The support of the class $K_c$ to the unknown gestural expression is evaluated by measuring the closeness between $M$-dimensional representative of class $K_c$ [obtained from step (5)] and the $M$-dimensional membership vector of the extracted feature set to belong to class $K_c$ [obtained from step (6)]. Less the difference, more is the support of class $K_c$ to the unknown pathological disorder.

8. Finally the support of all classes are evaluated for $c = [1, C]$. The class with the highest measure of the support is declared as the pathological disorder class of the unknown gestural expression.

It is noteworthy that the proposed T2 gesture induced pathological disorder recognition scheme takes care of both the inter- and intra-subject level uncertainty with the help of IVFS and GT2FS, and thus offers high classification accuracy. We have undertaken a comparative study of the proposed T2FS algorithm with seven state-of-the-art algorithms for the gesture induced pathological disorder recognition. The efficacy of all the eight contender algorithms to (recognize an unknown pathological disorder) is scrutinized with respect to three datasets of Jadavpur University research scholars of age (i) 25–30 years, (ii) 30–35 years, and (iii) 35–40 years. The relative performance of all the algorithms has been compared on the basis of five performance metrics—recall, precision, accuracy, F1_score and average error rate. The quality performance of the proposed T2FS algorithm is substantiated by the reported simulation results, with an accuracy of 83.95, 84.55, and 93.41% for the three respective datasets. The experimental study clearly reveals that T2FS (comprising attractive features of IVFS and GT2FS) outperforms its competitor algorithms with respect to five standard metrics irrespective of the datasets used.

Statistical significance of the results has been ascertained by four non-parametric tests including the McNemar's test, the Friedman test, the Iman-Davenport statistic, and the Bonferroni–Dunn post hoc analysis. It is worth mentioning that all the statistical tests affirm the rejection of the null hypothesis, portraying significant difference in the performance of different algorithms for the gesture driven physical disorder recognition. Moreover, T2FS based physical disorder recognition strategy turns out to be the conqueror achieving the highest average Friedman rank. It is further validated by the Bonferroni-Dunn test that, apart from the traditional IVFS and T1FS, the remaining five algorithms are outperformed by T2FS in a statistically significant manner.

## 5.7  Matlab Codes

```
% Code is written by Pratyusha Rakshit
% Under the guidance of Prof. Amit Konar
x=[0:0.001:10];
a=zeros(1,size(x,2));
b=a;
c=sort(repmat(2,1,20)+rand(1,20).*(repmat(4,1,20)-repmat(2,1,20)));
d=sort(repmat(4,1,20)+(1,20).*(repmat(6,1,20)-repmat(4,1,20)));
for i=1:10
    y(i,:)=trapmf(x,[a(i) b(i) c(i) d(i)]);
end;
```

```
plot(x,y,'LineWidth',2);
hold on;
UMF=max(y);
LMF=min(y);
plot(x,UMF,'k','LineWidth',4);
hold on;
plot(x,LMF,'k','LineWidth',4);
hold on;
```

**Results**: Sample run results are given in Table 5.1.

```
x=[0:0.001:0.1];
subect=10;
a=zeros(1,size(x,2));
b=a;
c=floor((repmat(0.015,1,subect)+rand(1,subect).*(repmat(0.045,1,-
subect)-repmat(0.015,1,subect)))*1000)/1000;
d=floor((repmat(0.045,1,subect)+rand(1,subect).*(repmat(0.075,1,-
subect)-repmat(0.045,1,subect)))*1000)/1000;
for i=1:subect
    mf(i,:)=trapmf(x,[a(i) b(i) c(i) d(i)]);
end;
figure();
plot(x,mf,'LineWidth',2);
hold on;
UMF=max(mf);
LMF=min(mf);
plot(x,UMF,'r','LineWidth',10);
hold on;
plot(x,LMF,'b','LineWidth',10);
grid on;
mf=[];
mf(1,:)=UMF;
mf(2,:)=LMF;
high(1)=max(mf(1,:));
low(1)=min(mf(1,:));
high(2)=max(mf(2,:));
low(2)=min(mf(2,:));
R=size(x,2);
S=zeros(size(mf,1),size(mf,2));
for n=1:size(mf,1)
    disp(n);
    F=[];
    for i=1:R
```

```
   if mf(n,i)==high(n) |mf(n,i)==low(n)
      S(n,i)=1;
      F=[F, x(i)];
   end;
end;
for i=1:R
   d(i)=abs(x(i)-F(1));
   for j=1:size(F,2)
      if abs(x(i)-F(j))<d(i)
         d(i)=abs(x(i)-F(j));
      end;
   end;
end;
[maxdist index]=max(d);
S(n,i)=0.3;
alpha=-log(0.3)/maxdist;
for i=1:R
   S(n,i)=exp(-alpha*d(i));
end;
figure();
surf([x;x],[mf(n,:);mf(n,:)],[S(n,:);zeros(1,size(x,2))]);
   colormap jet;
end;
```

**Results**: Sample run results are given in Fig. 5.6b.

# References

1. M. Quante, S. Hille, M.D. Schofer, J. Lorenz, M. Hauck, Noxious counterirritation in patients with advanced osteoarthritis of the knee reduces MCC but not SII pain generators: a combined use of MEG and EEG. J. Pain Res. **1**, 1 (2008)
2. M. Nixon, A.S. Aguado, Feature Extraction & Image Processing (Academic Press, London, 2008)
3. A. Konar, A. Chakraborty, Emotion Recognition: A Pattern Analysis Approach (Wiley, London, 2014
4. M. Parajuli, D. Tran, W. Ma, D. Sharma, Senior health monitoring using Kinect. in Commun Electron (ICCE) 2012 Fourth International Conference on IEEE, 2012, pp. 309–312
5. T.-L. Le, M.-Q. Nguyen, T.-T.-M. Nguyen, Human posture recognition using human skeleton provided by Kinect, in: Comput. Manag. Telecommun. (ComManTel), 2013 International Conference on IEEE, 2013, pp. 340–345
6. R.A. Clark, Y.-H. Pua, K. Fortin, C. Ritchie, K.E. Webster, L. Denehy et al., Validity of the Microsoft Kinect for assessment of postural control. Gait Posture. **36**, 372–377 (2012)
7. M. Oszust, M. Wysocki, Recognition of signed expressions observed by Kinect Sensor, in: Adv. Video Signal Based Surveill. (AVSS), 2013 10th IEEE International Conference on IEEE, 2013, pp. 220–225

8. X. Yu, L. Wu, Q. Liu, H. Zhou, Children tantrum behaviour analysis based on Kinect sensor, in: Intell. Vis. Surveill. (IVS), 2011 Third Chinese Conference on IEEE, 2011, pp. 49–52
9. C.C. Martin, D.C. Burkert, K.R. Choi, N.B. Wieczorek, P.M. McGregor, R.A. Herrmann, et al., A real-time ergonomic monitoring system using the Microsoft Kinect, in: Syst. Inf. Des. Symp. (SIEDS), 2012 IEEE, IEEE, 2012, pp. 50–55
10. N. Burba, M. Bolas, D.M. Krum, E.A. Suma, Unobtrusive measurement of subtle nonverbal behaviors with the Microsoft Kinect, in: Virtual Real. Work. (VR), 2012 IEEE, IEEE, 2012, pp. 1–4
11. K. Lai, J. Konrad, P. Ishwar, A gesture-driven computer interface using Kinect, in: Image Anal. Interpret. (SSIAI), 2012 IEEE Southwest Symposium on IEEE, 2012, pp. 185–188
12. B. Galna, G. Barry, D. Jackson, D. Mhiripiri, P. Olivier, L. Rochester, Accuracy of the Microsoft Kinect sensor for measuring movement in people with Parkinson's disease. Gait Posture, 1062–1068 (2014)
13. C. Metcalf, R. Robinson, A. Malpass, T. Bogle, T. Dell, C. Harris, et al., Markerless motion capture and measurement of hand kinematics: validation and application to home-based upper limb rehabilitation (2013)
14. E.E. Stone, M. Skubic, Fall detection in homes of older adults using the Microsoft Kinect. Biomed. Heal. Informatics, IEEE journal of biomedical and health informatics, $19(1)$, 290–301 (2015). doi:10.1109/JBHI.2014.2312180
15. D. Zhai, J.M. Mendel, Uncertainty measures for general type-2 fuzzy sets. Inf. Sci. (Ny) $181$, 503–518 (2011)
16. J.M. Mendel, R.I. John, F. Liu, Interval type-2 fuzzy logic systems made simple. Fuzzy Syst. IEEE Trans. $14$, 808–821 (2006)
17. L.A. Zadeh, Fuzzy sets. Inf. Control $8$, 338–353 (1965)
18. L.A. Zadeh, The concept of a linguistic variable and its application to approximate reasoning—I. Inf. Sci. (NY) $8$, 199–249 (1975)
19. A. Halder, A. Konar, R. Mandal, A. Chakraborty, P. Bhowmik, N.R. Pal, et al., General and interval type-2 fuzzy face-space approach to emotion recognition. IEEE Trans. Syst. Man Cybern. Syst. $43(3)$, 587–605 (2013)
20. J.M. Mendel, R.I.B. John, Type-2 fuzzy sets made simple. Fuzzy Syst. IEEE Trans. $10$, 117–127 (2002)
21. C. Wagner, H. Hagras, Toward general type-2 fuzzy logic systems based on zSlices. Fuzzy Syst. IEEE Trans. $18$, 637–660 (2010)
22. M. Biglarbegian, J.M. Mendel, On the justification to use a novel simplified interval type-2 fuzzy logic system, J. Intell. Fuzzy Syst. (n.d.)
23. M. Hao, J.M. Mendel, Similarity measures for general type-2 fuzzy sets based on the $\alpha$-plane representation. Inf. Sci. (Ny) $277$, 197–215 (2014)
24. J.M. Mendel, General type-2 fuzzy logic systems made simple: a tutorial. Fuzzy Syst. IEEE Trans. $22$, 1162–1182 (2014)
25. M.R. Rajati, J.M. Mendel, On Computing normalized interval type-2 fuzzy sets. IEEE Trans. Fuzzy Syst. (2013)
26. J.M. Mendel, D. Wu, Determining interval type-2 fuzzy set models for words using data collected from one subject: Person FOUs, in: Fuzzy Syst. (FUZZ-IEEE), 2014 IEEE International Conference on IEEE, 2014, pp. 768–775
27. D. Wu, J.M. Mendel, Designing practical interval type-2 fuzzy logic systems made simple, in: Fuzzy Syst. (FUZZ-IEEE), 2014 IEEE International Conference on IEEE, 2014, pp. 800–807
28. D.-J. Kim, S.-W. Lee, Z. Bien, Facial emotional expression recognition with soft computing techniques, in: Fuzzy Syst. 2005. FUZZ'05. 14th IEEE International Conference on IEEE, 2005, pp. 737–742
29. J. Mendel, H. Hagras, W.-W. Tan, W.W. Melek, H. Ying, Introduction to Type-2 Fuzzy Logic Control: Theory and Applications (Wiley, London, 2014)
30. Mendel, J., Wu, D. *Perceptual computing: aiding people in making subjective judgments* (vol 13, Wiley, London, 2010)

31. J.M. Mendel, Uncertain rule-based fuzzy logic system: introduction and new directions, (Prentice Hall PTR, Upper Saddle River, 2001)
32. H. Bustince, E.Barrenechea, M. Pagola, J. Fernandez, Z. Xu, B. Bedregal, J. Montero, H. Hagras, F. Herrera, B. De Baets. A historical account of types of fuzzy sets and their relationships. IEEE Trans. Fuzzy Syst. (2015)
33. J.M. Mendel, On type 2 fuzzy sets as granular models for words. Handb. Granul. Comput. 553–574 (2008)
34. D. Wu, J.M. Mendel, S. Coupland, Enhanced interval approach for encoding words into interval type-2 fuzzy sets and its convergence analysis. Fuzzy Syst. IEEE Trans. **20**, 499–513 (2012)
35. S. Coupland, J.M. Mendel, D. Wu, Enhanced interval approach for encoding words into interval type-2 fuzzy sets and convergence of the word fous, in: Fuzzy Syst. (FUZZ), 2010 IEEE International Conference on IEEE, 2010, pp. 1–8
36. R.E. Walpole, R.H. Myers, S.L. Myers, K. Ye, Probability and statistics for engineers and scientists (Macmillan New York, 1993)
37. P. Rakshit, A. Chakraborty, A. Konar, A.K. Nagar, Secondary membership evaluation in Generalized Type-2 Fuzzy Sets by evolutionary optimization algorithm, in: Fuzzy Syst. (FUZZ), 2013 IEEE International Conference on IEEE, 2013, pp. 1–8
38. Z. Zhang, Microsoft kinect sensor and its effect. Multimedia, IEEE. **19**, 4–10 (2012)
39. J. Tong, J. Zhou, L. Liu, Z. Pan, H. Yan, Scanning 3d full human bodies using kinects. Vis. Comput. Graph. IEEE Trans. **18**, 643–650 (2012)
40. S. Saha, M. Pal, A. Konar, R. Janarthanan, Neural Network Based Gesture Recognition for Elderly Health Care Using Kinect Sensor, in: Swarm, Evolutionary, and Memetic Computing, (Springer, Berlin, 2013), pp. 376–386
41. E. Stone, M. Skubic, Fall detection in homes of older adults using the Microsoft Kinect. IEEE journal of biomedical and health informatics (2014)
42. C.L. Lisetti, D.E. Rumelhart, Facial Expression Recognition Using a Neural Network., in: FLAIRS Conference, 1998, pp. 328–332
43. T.G. Dietterich, Approximate statistical tests for comparing supervised classification learning algorithms. Neural Comput. **10**, 1895–1923 (1998)
44. P.J. Bickel, Doksum, K. A, Mathematical statistics: basic ideas and selected topics. vol. 2, CRC Press, (2015)
45. S. García, D. Molina, M. Lozano, F. Herrera, A study on the use of non-parametric tests for analyzing the evolutionary algorithms' behaviour: a case study on the CEC'2005 special session on real parameter optimization. J. Heuristics **15**, 617–644 (2009)
46. J.H. Zar, Biostatistical analysis, Pearson Education India, (1999)
47. P. Rakshit, A. Konar, S. Das, L.C. Jain, A.K. Nagar, uncertainty management in differential evolution induced multiobjective optimization in presence of measurement noise. Syst. Man, Cybern. Syst. IEEE Trans. **1**, 922–937 (2013). doi:10.1109/TSMC.2013.2282118

# Chapter 6
# Probabilistic Neural Network Based Dance Gesture Recognition

With the growing interest in the domain of human computer interaction these days, budding research professionals are coming up with novel ideas of developing more versatile and flexible modes of communication between a man and a machine. Using the attributes of internet, the scientists have been able to create a web based social platform for learning any desired art by the subject himself/herself, and this particular procedure is termed as electronic learning or e-learning. In this chapter, we propose a novel application of gesture dependent e-learning of dance. This e-learning procedure may provide help to many dance enthusiasts who cannot learn the art because of scarcity of resources despite having great zeal. The chapter mainly deals with recognition of different dance gestures of a trained user such that after detecting the discrepancies between the gestures shown and actually performed by a novice; the user can rectify his faults. The elementary knowledge of geometry has been employed to introduce the concept of planes in the feature extraction stage. Actually, five planes have been constructed to signify major body parts while keeping the synchronous parts in one unit. Then four distances and four angular features have been obtained to provide entire positional information of the different body joints. Finally, using a probabilistic neural network the dance gestures have been classified after training the said network with sufficient amount of data recorded from numerous subjects to maintain generality. To check the capability of the discussed method, it has been compared with various standard classifiers in terms of performance indices and in each case the proposed framework has surpassed or provided nearly equal performance as given by the other networks.

Contributed by Sriparna Saha, Rimita Lahiri and Amit Konar

© Springer International Publishing AG 2018
A. Konar and S. Saha, *Gesture Recognition*, Studies in Computational
Intelligence 724, DOI 10.1007/978-3-319-62212-5_6

## 6.1   Introduction

With the technological advancement in the field of human computer interaction (HCI), it has become quite easy to develop interactive models that provide effective ways of communication in terms of cost efficiency, time complexity and portability as well. Using the newly developed sensors with RGB camera embedded within the device itself, it is possible to detect different human body gestures and analyse the same by generating skeletal data, containing three dimensional body joints information in terms of position as well as orientation. Literally, the term 'gesture' signifies the different postures that a human structure embodies while displaying various actions as well as emotions. Dance is considered as one of the best ways of conveying human emotions. Instead of considering static uncorrelated postures of a user at different time instances to convey his mental states, dance is chosen as a preferred alternative as it incorporates both facial expressions and body movements, which collectively enhances the ability of a user to convey his emotions. Moreover, with the smooth transition between different dance poses, it adds an aesthetic value which enables a layman to understand the message that is intended to be conveyed through that concerned act.

As an obvious consequence of such rapid progress of different engineering methodologies, social networking is a part and parcel of our daily life. This motivated researchers to delve further in the concerned domain and come up with ideas for building an immersive and collaborative environment to support more realistic and versatile interactions between human beings and a computer. To engage a larger percentage of the research fraternity in solving these real time problems, many popular multinational research organizations are coming up with innovative challenges. For example, Huawei 3DLife/EMC2 challenge [1] provides a platform for demonstrations of online human computer interactive systems and thus encourages the research community to solve the problems faced by the industry or improvise the already acquired solutions to move closer to the desired output. With the recent advent of internet, e-learning program has emerged as one of the most popular internet accessory, which helps a certain human being to acquire different skills online with the help of internet, which wouldn't have been possible for the user to learn in person otherwise. It is needless to tell, machine intelligence plays an important role in these kinds of systems. In this chapter, we propose a novel cost effective and fast e-learning procedure of detecting different dance postures embodied by an already trained subject, so that a novice can be able to learn from those postures and correct himself in case of any wrong gesture.

Several methodologies have been adopted in the said domain, but none of them has been able to provide optimal results for e-learning of dance gestures. This motivated us to enlighten the other drawbacks and improvise the methodologies as required. This chapter has emphasized on mainly five different classical dance gestures corresponding to 'anger', 'fear', 'happiness', 'sadness' and 'relaxation'. For further analysis of the above mentioned five key emotive gestures, feature extraction seems to be an area of utmost importance in skeleton based system analysis because inaccurate feature selection may have an adverse impact upon the system performance as a whole. Here, the Microsoft's Kinect sensor [1, 2] plays a key role of registering the three dimensional RGB images, storing depth map at a predetermined frame rate and generating three dimensional positional information in terms of coordinates from a skeleton comprising of twenty different human body joints with the help of software development kit (SDK). An innovative feature space has been formulated using a plane based approach and the concerned feature set contains both positional and angular parameters to avoid unnecessary loss of information. Since hand and leg movements need to be focused on while analyzing any dance gesture, the feature set has been constructed such that it includes everyinformation corresponding to hand and leg joints without loss of generality. In this chapter, the elementary geometric knowledge of creating planes and deriving perpendicular distances has been used. Five different planes have been created including a reference plane based on body core (formed by spine, hip center and shoulder center joints), as this part is relatively stationary while performing any dance movement and the other four functioning planes have been constructed corresponding to the arm and leg joints respectively. In the next step, the orientation of the planes corresponding to the concerned arm and leg joints have been derived with respect to the reference plane and the perpendicular distance from the spine to each of the plane is calculated while conducting the experiment for each subject, yielding a feature set comprising of four distance based and four orientation based features for each frame of each subject. Further, using a probabilistic neural network (PNN) algorithm [3–6], an appreciable matching accuracy of 89.7% has been attained. The proposed framework has not only generated better performance in terms of precision parameters but it has improved the hardware and time complexity by a noticeable extent. Although the proposed system has been implemented only for Indian classical dance postures, but it can be applicable for any dance form with clearly distinguishable gestures.

This chapter is divided into four sections. The detailed literature survey is provided in Sect. 6.2. The overview of proposed framework along with the preliminary concepts has been discussed in Sects. 6.3 and 6.4. Section 6.5 presents the experimental results obtained during different stages of the execution of the proposed model. Section 6.6 concludes the discussion by combining our findings.

## 6.2   Previous Work in the Field of Gesture Recognition

Literature show much research has already been done in the said area, but still remains a discrepancy between the real world scenario and the artificially framed simulated environment, and our aim remains to minimize that gap as far as possible. Saha et al. explored a completely different aspect of the same problem by presenting a novel gesture recognition algorithm to segregate between different emotive gestures of Indian classical dance [2]. But the works lacks an efficient feature extraction stage, which is overcome in the proposed framework. Waldherr et al. developed an application of human robot interaction by employing camera based human movement tracking approach in order to control the movement of a robotic arm [7]. Essid et al. presented a new idea of multi modal dance collection based approach that facilitates the interaction between human beings in virtually framed environment [8]. As a response to the Huawei 3DLife challenge, Alexiadis et al. proposed a Kinect based approach for creating an online dance evaluating software, which measures the worth of a performance with respect to a standard one and generates feedback for further improvement [1].

Xu et al. proposed a new feed forward neural network based approach to detect and recognize hand gestures in order to develop a virtually occurring driving training system called Self Propelled Gun (SPG) [9]. Murakami et al. extended the neural network based approach and used another variant of neural network named recurrent neural network to carry out finger alphabet recognition dynamically [10]. The above mentioned method has implemented the framework for a small vocabulary, but there is further scope of improvement in terms of extent of application of the designed framework. Mao et al. proposed a new algorithm that iteratively determines the structure of a probabilistic neural network (PNN) by recurrently calculating the value of a smoothing parameter using a genetic algorithm (GA) technique [11]. Wu et al. developed a novel leaf pattern recognition algorithm, used for plant classification purpose, by employing PNN [3].

In [12], stick figure diagrams have been used to sequence ballet dance postures. Here human motion sensor device is used. In this proposed work we are also using the Kinect sensor, which is a human motion sensor device. Reference [13] explains about an algorithm, which resorts to the placing of two cameras orthogonally at a mid-body height for dance gesture recognition. But here we are using only one camera, thus our work becomes more useful, as it is difficult to place two cameras at the middle point of the body. The whole arrangement made in [13] results in a technique, which is inconvenient for e-learning purposes. In [14], single and multiple gestures are used to create dance datasets, which are classified by using hidden Markov model. Multiple calibrated cameras are required to project the dance gestures three dimensionally. These cameras are very costly and hence are not suitable for the purpose of e-learning. In this work a single camera is sufficient for carrying out the entire procedure. Reference [15] removes ambiguity in posture recognition by using Markov model and K-means algorithm. Background image is of immense importance for posture recognition. Error in the output is contributed by noisy background, improper texture of the dress and an inappropriate distance of the dancer from the camera. The above mentioned problems are successfully dealt with in [16] by the usage of the proposed algorithm where histograms are used. In [17], active contours and neural networks are used to track primary human postures like sitting, bending and squatting. Here background subtraction technique has been used at the preprocessing stage. Another approach is the application of a decision tree for human posture recognition which is discussed at length in [18]. Information gain can be measured by entropy which is the basis of decision tree [19]. A genetic fuzzy finite state machine is applied to a fuzzy theoretical approach to posture recognition. Here inputs to the state machine are provided and fuzzy inference rules are used to obtain the outputs [20]. Reference [21] uses comparison of supervised and unsupervised learning classifiers for recognizing human postures.

Dance performance is normally accompanied with melody [22]. Presently a major goal of the computer graphics research and the robotics research is the creation of a realistic motion sequence. Dynamic simulation is a well-known method which is often used. First, a programmer creates a proper motion sequence manually, and then, the sequence is optimized in order to maintain balance over the whole motion sequence by using a dynamics simulator. This is a useful technique but is quite tedious. There is another popular method which is based on the idea of motion capture system. This technique consists of 'Motion Graph' [23–25], which is the application of statistics to a set of motion capture data to synthesize new motion sequences [26]. However, most of these techniques do not consider human emotion and recognition despite the fact that humans often exhibit behaviors based on emotion and recognition. There are a few previous methods to consider human emotion when creating motion sequences. Kim et al. [27] proposed a motion enhancement method that could analyze the rhythm of motion and synthesize new motions based on the results of motion rhythm analysis, and by using this method, we could synthesize rhythmic motions. However, for example, for creating new dance performance, the music signal needed to be synchronized to the rhythm of motion. Stone et al. [28] propose a method that could synthesize new utterance performance synchronized to an input speech signal. However, this method needed considerable manual processing, was not automatically, and required a great deal of time. Reference [29] discusses some works on traditional Korean traditional dance forms. Based on these representations of Korean emotions various conclusions have been drawn on geographical, religious, and socio-cultural background of the place. Among these, 'Han' is the characteristic emotion.

## 6.3 Russell's Circumplex Model

In this chapter we have worked on recognizing five types of gestures. They are 'anger', 'fear', 'happiness', 'sadness', and 'relaxation'. While 'anger' is expressed by aggressive hand movements, 'relaxation' shows static hand postures but kept at a definite angle. The motion analysis has been carried out by the Kinect sensor, which tracks the skeleton of the dancer, while performing.

The feature extraction process in this chapter is based on Russell's Circumplex model [30]. The Circumplex model of emotion was first established by James Russell [31, 32]. This model provides that emotions are distributed in a two-dimensional circular space, containing arousal and valence dimensions. Arousal represents the vertical axis and valence is in the direction of horizontal axis, while the center of the circle represents a neutral valence and a medium level of arousal. In this model, emotional states can be presented at any level of valence and arousal, or at a neutral level of one or both of these factors. Circumplex models have been implemented most commonly to test stimuli of emotion words, emotional facial expressions, and affective states. Russell and Lisa Feldman Barrett describe their modified Circumplex model as representative of core affect, or the

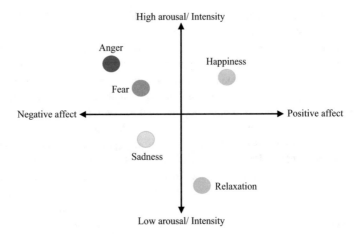

**Fig. 6.1** Model for gesture classification based on Circumplex model

most elementary feelings that are not necessarily directed toward anything. Different prototypical emotional episodes, or clear emotions that are evoked or directed by specific objects, can be plotted on the Circumplex, according to their levels of arousal and pleasure.

The model shown in Fig. 6.1 gives an idea about the classification of gestures based on type and intensity. According to it, if 'Happiness' and 'Relaxation' are positive gestures then 'Anger', 'Fear' and 'Sadness' are negative gestures. On the other hand, while 'Anger', 'Fear' and 'Happiness' gestures indicate higher arousal or intensity, 'Relaxation' and 'Sadness' gestures show the opposite characteristics.

## 6.4   Overview of the Proposed Work

The steps required for the proposed work for e-learning of dance gestures are given in Fig. 6.2. The elaboration of the overall process is given in following sub-sections.

### 6.4.1   Skeleton Formation Using Microsoft's Kinect Sensor

Kinect sensor typically looks like a long bar like device with a set of IR (infrared) and RGB (red, green, blue) cameras embedded within it. The device was primarily developed as a game peripheral controller, but later it found extensive application in the domain of gesture recognition. Actually, the Kinect sensor generates three dimensional joint coordinates information by generating a skeletal image of the subject present in front of it within a threshold distance 1.2–3.5 m using SDK toolkit. The IR cameras are primarily responsible for sensing and thus generating depth map after processing the data recorded from the user standing in front of it.

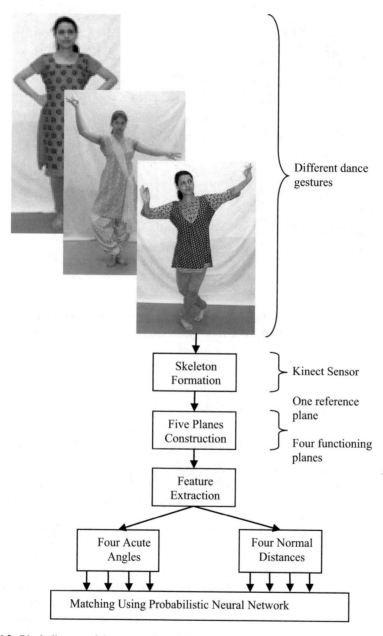

Fig. 6.2 Block diagram of the proposed work for e-learning of dance

A sample RGB image with its corresponding skeleton obtained using Kinect sensor is provided in Fig. 6.3. For the proposed work, the required 15 joints, which are hip center (*HC*), spine (*S*), shoulder center (*SC*), elbow left (*EL*), wrist left (*WL*),

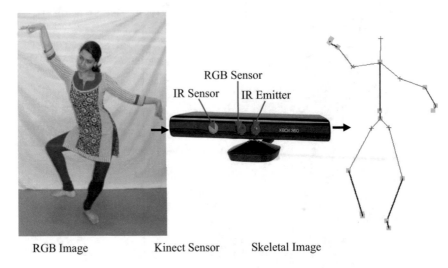

RGB Image                Kinect Sensor            Skeletal Image

**Fig. 6.3** Sample image to depict skeleton extraction procedure using Kinect sensor

hand left (*HL*), knee left (*KL*), ankle left (*AL*), foot left (*FL*), elbow right (*ER*), wrist right (*WR*), hand right (*HR*), knee right (*KR*), ankle right (*AR*), foot right (*FR*) are highlighted.

### 6.4.2   Five Planes Construction

**For the next step of feature extraction, we have created five planes.**

(a) *Plane 1 ($P_{ref}$)*: For body core—hip center (*HC*), spine (*S*), shoulder center (*SC*).
(b) *Plane 2 ($P_{LA}$)*: For left arm—elbow left (*EL*), wrist left (*WL*), hand left (*HL*).
(c) *Plane 3 ($P_{RA}$)*: For right arm—elbow right (*ER*), wrist right (*WR*), hand right (*HR*).
(d) *Plane 4 ($P_{LL}$)*: For left leg—knee left (*KL*), ankle left (*AL*), foot left (*FL*).
(e) *Plane 5 ($P_{RL}$)*: For right leg—knee right (*KR*), ankle right (*AR*), foot right (*FR*).

$P_{ref}$ is nearly constant for a specific gesture, thus it is taken as the reference plane. The rest of the planes ($P_{LA}$, $P_{RA}$, $P_{LL}$, $P_{RL}$) are use as functioning planes. Dance can be considered as an amalgamation of several rhythmic body movements that collectively expresses an art, conveying a story behind it. Since body movements are involved in this case, hence body parts obviously play an important role in this art. For transitional movements, the body parts majorly involved is the arms and the legs. The Kinect sensor provides three dimensional joint coordinates of three joints corresponding to each arm and leg as well. To accommodate all three joint information in a single entity a plane concept has been introduced, such that a single functioning plane can singlehandedly include all the joints information. The pictorial representation of these planes is given in Fig. 6.4.

**Fig. 6.4** Construction of
planes for skeletal image from
Fig. 6.3

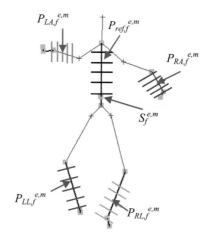

### 6.4.3  *Feature Extraction*

Let there are $M$ number of subjects and each subject is depicting $E$ different emotions
using body gestures through dance. To demonstrate each $e$ ($1 \leq e \leq E$) gesture,
we need $F$ number of frames and from each frame $f$ ($1 \leq f \leq F$), $R = 8$ features are
extracted. Now for a specific subject $m$ ($1 \leq m \leq M$) for a specific frame $f$ for a
specific emotion $e$, the extracted features are:

(a) *Feature 1 ($e_{f,1}^m$)*: Acute angle between $P_{ref,f}^{e,m}$ and $P_{LA,f}^{e,m}$.
(b) *Feature 2 ($e_{f,2}^m$)*: Acute angle between $P_{ref,f}^{e,m}$ and $P_{RA,f}^{e,m}$.
(c) *Feature 3 ($e_{f,3}^m$)*: Acute angle between $P_{ref,f}^{e,m}$ and $P_{LL,f}^{e,m}$.
(d) *Feature 4 ($e_{f,4}^m$)*: Acute angle between $P_{ref,f}^{e,m}$ and $P_{RL,f}^{e,m}$.
(e) *Feature 5 ($e_{f,5}^m$)*: Normal distance between $S_f^{e,m}$ and $P_{LA,f}^{e,m}$.
(f) *Feature 6 ($e_{f,6}^m$)*: Normal distance between $S_f^{e,m}$ and $P_{RA,f}^{e,m}$.
(g) *Feature 7 ($e_{f,7}^m$)*: Normal distance between $S_f^{e,m}$ and $P_{LL,f}^{e,m}$.
(h) *Feature 8 ($e_{f,8}^m$)*: Normal distance between $S_f^{e,m}$ and $P_{RL,f}^{e,m}$.

A natural question may arise regarding the reason behind the requirement of both
angular and distance based features. Actually, using only one sort of features, it is
not feasible to provide complete positional information about different body joints.
As position not only include translational motion but orientation is also very
important for dance gestures. While dancing, the body parts of a dancer do not
move following a linear displacement, those parts are also realigned during execution of different gestures. Hence it is very important to include both distance
based as well as orientation based features in order to register the extent of
realignment since that has quite an impact on the resulting output. For better
understanding, the formation of five planes corresponding to the skeletal image
from Fig. 6.3 is given in Fig. 6.4.

Let two normal vectors corresponding to two planes $p_1$ and $p_2$ are

$$\vec{n}_1 = (x_1, y_1, z_1) \tag{6.1}$$

$$\vec{n}_2 = (x_2, y_2, z_2) \tag{6.2}$$

Then the angle $\alpha$ between those two planes is calculated using (6.3).

$$\alpha(\vec{n}_1, \vec{n}_2) = \arccos \frac{|x_1 \times x_2 + y_1 \times y_2 + z_1 \times z_2|}{\sqrt{x_1^2 + y_1^2 + z_1^2} \times \sqrt{x_2^2 + y_2^2 + z_2^2}} \tag{6.3}$$

Let the equation of a plane $p$ is

$$ax + by + cz = d \tag{6.4}$$

also the co-ordinate of a point $S$ is $[A_1, B_1, C_1]$. Then the normal distance $nd$ from $S$ to plane $p$ is $nd$.

$$nd = \frac{aA_1 + bB_1 + cC_1 - d}{\sqrt{a^2 + b^2 + c^2}} \tag{6.5}$$

For better visualization Fig. 6.5 is provided which shows the procedure for acute angle and normal distance measurement. The pseudo code to calculate these features is given in Table 6.1.

The feature vector $(G_e^m)$ obtained for $e$th gesture for $m$th subject for all frames is given in (6.6).

$$G_e^m = \begin{bmatrix} e_{1,1}^m & e_{2,1}^m & \cdots & e_{f,4}^m & \cdots & e_{F-1,8}^m & e_{F,8}^m \end{bmatrix}^T \tag{6.6}$$

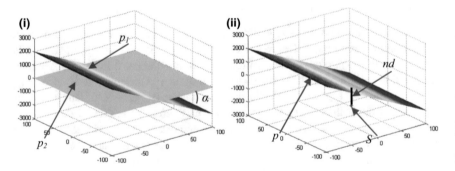

**Fig. 6.5** Procedure for calculation of acute angle and normal distance

**Table 6.1** Pseudo-code for feature extraction

Input: 3D joint co-ordinates to form reference plane: $J_{ref,1}$, $J_{ref,2}$ and $J_{ref,3}$
Any three 3D joint co-ordinates: $J_1$, $J_2$ and $J_3$
3D joint co-ordinate of spine $S$

**Output**: One acute angle feature say $\alpha$ and one normal distance say $nd$

**Procedure:**
**Begin**

$$n_{ref} = \mathrm{cross}\left(J_{ref,1} - J_{ref,2}, J_{ref,1} - J_{ref,3}\right)$$
$$d_{ref} = J_{ref,1}(1) \times n_{ref}(1) + J_{ref,1}(2) \times n_{ref}(2) + J_{ref,1}(3) \times n_{ref}(3)$$
$$n = \mathrm{cross}(J_1 - J_2, J_1 - J_3)$$
$$d = J_1(1) \times n(1) + J_1(2) \times n(2) + J_1(3) \times n(3)$$
$$flag_1 = ((n_{ref}(1)) \times (n(1))) + ((n_{ref}(2) \times n(2))) + ((n_{ref}(3) \times n(3)))$$
$$flag_2 = (((n_{ref}(1))^2) + ((n_{ref}(2))^2) + ((n_{ref}(3))^2))^{1/2}$$
$$flag_3 = (((n(1))^2) + ((n(2))^2) + ((n(3))^2))^{1/2}$$
$$\alpha = (\mathrm{arc}\cos(|flag_1|/(flag_2 \times flag_3))) \times 180/\pi$$
$$flag_4 = ((n(1)) \times (S(1))) + ((n(2)) \times (S(2))) + ((n(3)) \times (S(3))) - d$$
$$flag_5 = (((n(1))^2) + ((n(1))^2) + ((n(1))^2))$$
$$nd = |flag_4/flag_5|$$

**End**

## 6.4.4 Matching of Unknown Gesture Using Probabilistic Neural Network

Neural networks are often used for classification purpose in pattern recognition problems, different neural networks are based upon different learning rules. For example, the back propagation neural network (BPNN) employs the heuristic rule by modifying the parameters by small amounts at each iteration and thus improving the system performance. Hence BPNN takes a lot of time for training purpose, besides that it has a tendency that the neural weights get trapped at local minima. So instead of using back propagation network, in this case probabilistic neural network (PNN) [3–6] has been preferred over BPNN because of the former's ease of use and low time complexity.

A PNN is basically a mathematical implementation of a statistical algorithm termed as Kernel Discriminant Analysis, in which the different actions are arranged in a multilayered feed forward network (FFNN) with three layers as shown in Fig. 6.6. The structure of a PNN is very similar to a BPNN; only the sigmoid activation function has been replaced with a statistically derived exponential function in such a way that the resulting network assures optimal convergence to Bayes decision strategy as the dimension of the training set increases. The most prominent advantage of PNN over BPNN is that new training samples can be added without disturbing the already adapted weights of the existing neurons.

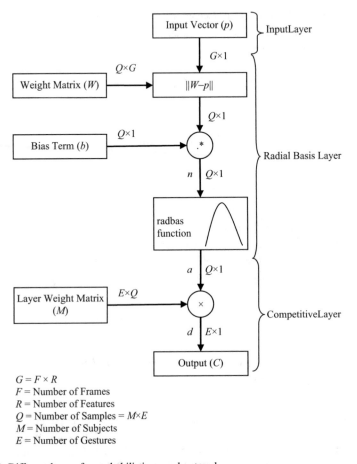

$G = F \times R$
$F$ = Number of Frames
$R$ = Number of Features
$Q$ = Number of Samples = $M \times E$
$M$ = Number of Subjects
$E$ = Number of Gestures

**Fig. 6.6** Different layers for probabilistic neural network

In this chapter, the PNN is comprised of three layers, an input layer, radial basis layer and a completive layer. Once an input vector is fed to the already trained network, the first layer computes the closeness of the fed input with respect to weight vectors in terms of distance measures, and those distances are scaled using the nonlinear radial basis function. Basically, the radial basis layer sums up the contributions of each class and generates a vector formed by probabilities and the competitive layer assigns the fed input pattern to the class having the highest probability. The network structure of PNN is as follows. Figure 6.6 provides a detailed overview of the network structure used for the current chapter.

(i) *Input Layer*: The input vector denoted by $p$ is of the dimension $G \times 1$ in this case $G = F \times R$. In this way, all the features of all the frame for a specific gesture corresponding to a particular subject has been accomodated in a single feature vector.

(j) *Radial Basis Layer*: The distances between the fed input vector and weight vectors of the trained network is determined. Here, a dot product measure has been adopted to signify distance with respect to the rows of weight matrix $W$ having dimension of $Q \times G$. So, a matrix of dimension $Q \times 1$ is formed such that $i$th element of the vector $\|W - p\|$ is actually the dot product of $i$th row of $W$ and the input vector. Here $Q$ is the number of samples taken in the training phase and G is same as the feature vector.

$$\|W - p\|_i = W(i, :).p \tag{6.7}$$

where $i$ denotes the $i$th term.

In the next step, an element wise multiplication is carried out between $\|W - p\|$ and the bias vector ($b$) of the same dimension as $\|W - p\|$ to generate a vector $n$ of dimension $Q \times 1$ using the equation,

$$n = \|W - p\|. * b \tag{6.8}$$

Further, each element of $a$ is replaced with its radial basis output. The radial basis function is defined as,

$$\text{radbas}(n) = e^{-n^2} \tag{6.9}$$

Finally, the output of radial basis layer is obtained as,

$$a = \text{radbas}(\|W - p\|. * b) \tag{6.10}$$

(k) *Competitive Layer*: Firstly, there is no bias term in the competitive layer. The output matrix of radial basis layer ($a$) is multiplied with a layer weight matrix ($M$) to generate a $E \times 1$ dimensional vector $d$, depending upon which a competitive function (func) assigns 1 to the largest value of $d$. $M$ is a matrix of $E \times Q$ dimension containing the class information of the training samples. Basically, it is a sparse matrix comprising of target vectors with one 1 in each column. Depending upon the class label, 1 is assigned to the corresponding row of the particular sample column. In this case the index 1 suggests the number of samples classified by the network. To obtain the output vector in a properly organized structure there are several functions that can transform it into desired form in different languages. Finally the output of the competitive layer $C$ is of $E \times 1$ dimension with 1 assigned to corresponding class. Suppose the input vector belongs to $e$th class then the $e$th element of $C$ is assigned 1 and others are assigned as 0.

## 6.5   Experimental Results

Here for these proposed work, we have taken data from forty subjects (i.e., $M = 40$). Each subject is asked to perform five gestures (i.e., $E = 5$) each showing one emotion (e.g. 'anger', 'fear', 'happiness', 'sadness' and 'relaxation') through Indian classical dance. Thus total number of samples, $Q = M \times E = 40 \times 5 = 200$. Each dance video is consisting of 4 s duration. As Kinect sensor has sampling frequency rate is 30fps, thus $F = 4 \times 30 = 120$. The number of rows in feature vector $G_e^m$ is $F \times R = 120 \times 8 = 960$ as number of features $R = 8$. Thus the dimension of $G$ is $960 \times 1$.

The acquired RGB and skeletal images from Kinect sensor with corresponding feature space for five gestures for 25th subject is given in Table 6.2. As it is clearly seen from Table 6.3, the starting gestures of all the videos are almost same. Table 6.4 shows the confusion matrix for the proposed system for all the five gestures considered in this chapter.

Along with classification accuracy, misclassification rate should also be checked with equal importance while evaluation of the performance of a classifier. Table 6.4 presents a confusion matrix of the five classes taken under consideration. The large values at the diagonal positions signify high accuracy with significantly lower misclassification rates.

We have validated our proposed work's performance with back propagation neural network (BPNN) [33], recurrent neural network (RNN) [10], feed forward neural network (FFNN) [9], ensemble classifier using binary tree (ECBT) [34], linear support vector machine (LSVM) [34], support vector machine with radial basis function (RBFSVM) [2], k-nearest neighbor (kNN). Neural networks are

**Table 6.2**  Pseudo-code for probabilistic neural network

| |
|---|
| **Input**: Number of training samples $Q$, Weight matrix $W$, Input vector $p$, Bias $b$, Number of classes $E$ |
| **Output**: Each test gesture is assigned a class label in the vector $C$ of dimension $E \times 1$ |
| **Procedure:**<br>**Begin**<br>**For** $i = 0$ to $i <= Q$<br>$\lVert W - p \rVert [i] = W[i].\, p$<br>**End.**<br>$b = 1$<br>$n = \lVert W - p \rVert . * b$<br>$a = exp\,(-\,n^2)$<br>Calculate $M$ if $i^{th}$ sample belongs to $j^{th}$ class then 1 is assigned in $j^{th}$ row and $i^{th}$ column.<br>$d = M \times a$<br>Calculate $max(d)$ and 1 is assigned to the corresponding index indicating the class label, generating class label matrix $C$ of dimension $E \times 1$<br>**End.** |

**Table 6.3** Results obtained from subject number 25

| Gesture type | Para-meters | Frame No. 20 | Frame No. 60 | Frame No. 100 |
|---|---|---|---|---|
| Anger | RGB Image | | | |
| | Skeletal Image | | | |

(continued)

**Table 6.3** (continued)

| Gesture type | Para-meters | Frame No. 20 | Frame No. 60 | Frame No. 100 |
|---|---|---|---|---|
| | Features | 43.137 | 36.122 | 43.710 |
| | | 48.299 | 84.639 | 21.098 |
| | | 7.519 | 8.566 | 9.697 |
| | | 6.309 | 58.871 | 70.102 |
| | | 0.097 | 0.131 | 0.060 |
| | | 0.103 | 1.817 | 0.088 |
| | | 0.014 | 0.097 | 0.116 |
| | | 0.032 | 0.176 | 0.067 |
| Fear | RGB Imag | | | |

(continued)

**Table 6.3** (continued)

| Gesture type | Para-meters | Frame No. 20 | Frame No. 60 | Frame No. 100 |
|---|---|---|---|---|
| | Skeletal Image | | | |
| | Features | 43.208 | 53.377 | 56.810 |
| | | 45.853 | 34.056 | 31.081 |
| | | 5.451 | 12.129 | 64.756 |
| | | 6.319 | 14.143 | 16.067 |
| | | 0.096 | 0.556 | 0.288 |
| | | 0.103 | 0.096 | 0.083 |
| | | 0.029 | 0.135 | 0.454 |
| | | 0.032 | 0.095 | 0.088 |

(continued)

**Table 6.3** (continued)

| Gesture type | Para-meters | Frame No. 20 | Frame No. 60 | Frame No. 100 |
|---|---|---|---|---|
| Happ-iness | RGB Image | | | |
| | Skeletal Image | | | |

(continued)

**Table 6.3** (continued)

| Gesture type | Para-meters | Frame No. 20 | Frame No. 60 | Frame No. 100 |
|---|---|---|---|---|
| | Features | 43.137<br>45.106<br>6.123<br>7.317<br>0.096<br>0.103<br>0.024<br>0.032 | 41.341<br>69.122<br>19.586<br>14.253<br>0.296<br>0.368<br>0.149<br>0.125 | 48.293<br>57.917<br>31.005<br>15.996<br>0.306<br>0.281<br>0.232<br>0.119 |
| Sad-ness | RGB Image | | | |

(continued)

Table 6.3 (continued)

| Gesture type | Para-meters | Frame No. 20 | Frame No. 60 | Frame No. 100 |
|---|---|---|---|---|
| | Skeletal Image | | | |
| | Features | 43.083 | 56.517 | 54.436 |
| | | 41.159 | 77.769 | 75.685 |
| | | 6.414 | 21.857 | 24.690 |
| | | 6.305 | 60.849 | 38.096 |
| | | 0.097 | 0.128 | 2.516 |
| | | 0.102 | 0.301 | 1.809 |
| | | 0.021 | 0.130 | 0.224 |
| | | 0.032 | 0.023 | 0.200 |

(continued)

**Table 6.3** (continued)

| Gesture type | Para-meters | Frame No. 20 | Frame No. 60 | Frame No. 100 |
|---|---|---|---|---|
| Relax-ation | RGB Image | | | |
| | Skeletal Image | | | |

(continued)

**Table 6.3** (continued)

| Gesture type | Para-meters | Frame No. 20 | Frame No. 60 | Frame No. 100 |
|---|---|---|---|---|
| | Features | 43.024 | 59.213 | 60.313 |
| | | 4.011 | 59.390 | 63.818 |
| | | 5.992 | 69.337 | 68.453 |
| | | 6.287 | 66.884 | 70.384 |
| | | 0.097 | 0.107 | 0.128 |
| | | 0.101 | 0.210 | 0.136 |
| | | 0.033 | 0.004 | 1.002 |
| | | 0.032 | 0.040 | 0.104 |

**Table 6.4** Confusion matrix for the probabilistic neural network

| Actual Class | Predicted Class | | | | |
|---|---|---|---|---|---|
| | | Anger | Fear | Happiness | Relaxation | Sadness |
| | Anger | 89.792 | 3.984 | 2.561 | 1.874 | 1.819 |
| | Fear | 2.311 | 90.517 | 3.549 | 1.195 | 2.428 |
| | Happiness | 3.058 | 2.329 | 89.911 | 2.079 | 2.623 |
| | Relaxation | 1.899 | 2.718 | 1.235 | 91.437 | 2.711 |
| | Sadness | 1.567 | 1.934 | 2.539 | 3.877 | 90.083 |

biologically inspired classification algorithms. The main building blocks of such algorithms are simply neuron like processing units termed as nodes, arranged in layers. Connections between nodes of adjacent layers are denoted in terms of weights which are updated iteratively during the learning phase. ECBT aims to classify multiple classes by dividing the original dataset into binary class subsets and developing a binary model for each such newly formed subset. This algorithm is based on the principal of adaptive boosting (AdaBoost) technology. LSVM attempts to classify simply constructing hyper-plane, while RBFSVM, another variant of SVM, that classifies the data by designing a non-linear Gaussian radial basis function based kernel such that the resulting algorithms easily fits into a maximum margin hyper-plane in a transformed feature space. kNN belongs to the category of instance based learning, where the outputs are generated in terms of membership, such that a data sample is assigned a class label depending upon the votes of its 'k' nearest neighbors after employing majority voting strategy. The performance metrics include accuracy, precision, sensitivity and specificity. The comparison results for these metrics are given in Table 6.5.

**Table 6.5** Comparison of proposed work with existing literatures

| Algorithms | Accuracy | Precision | Sensitivity | Specificity | |
|---|---|---|---|---|---|
| PNN | 0.917 (0.077) | 0.904 (0.035) | 0.926 (0.011) | 0.925 (0.030) | $t$ |
| BPNN | 0.883 (0.095) | 0.891 (0.080) | 0.908 (0.031) | 0.895 (0.030) | 0.273 (+) |
| RNN | 0.866 (0.036) | 0.878 (0.042) | 0.877 (0.031) | 0.891 (0.098) | 0.584 (+) |
| FFNN | 0.827 (0.062) | 0.812 (0.037) | 0.829 (0.050) | 0.834 (0.084) | 0.899 (+) |
| ECBT | 0.850 (0.093) | 0.845 (0.066) | 0.858 (0.021) | 0.833 (0.029) | 0.549 (+) |
| LSVM | 0.726 (0.056) | 0.756 (0.093) | 0.715 (0.088) | 0.757 (0.085) | 1.977 (+) |
| RBFSVM | 0.789 (0.041) | 0.793 (0.086) | 0.803 (0.012) | 0.798 (0.079) | 1.443 (+) |
| kNN | 0.747 (0.068) | 0.763 (0.0684) | 0.754 (0.030) | 0.748 (0.041) | 1.636 (+) |

We have used paired *t*-test for statistical comparison where PNN acts as the reference classifier and it is compared with accuracy values of other six standard classifier results given in Table 6.5.

$$t = \frac{\vec{x}_1 - \vec{x}_2}{\sqrt{\left((\sigma_1^2 + \sigma_2^2)/2\right)}} \tag{6.11}$$

where $\vec{x}_1$ and $\sigma_1^2$ are the mean and standard deviation of the proposed algorithm, $\vec{x}_2$ and $\sigma_1^2$ are the same for any one out of the six competitive algorithms.

For all the cases '+' significance is obtained which indicates that *t* value of 49 degrees of freedom is significant at a 0.05 level by two-tailed *t*-test. For this test we have taken into account the accuracy values.

## 6.6   Conclusion

This chapter provides a cost effective and easy to use way of e-learning of dance, the experiment has been implemented over Indian classical dance gestures only, but it is applicable to learning of any kind of dance form. While conducting the experiments, maximum accuracy obtained is 91.7% which is fairly good in this domain. Moreover, the feature extraction strategy includes introduction of planes passing through body joints. The feature extraction strategy adopted in the present work is very efficient and has little scope of error, which reduces the error complexity of the entire system as a whole. Employing this strategy has not only improved the performance of the proposed framework, but it has also become easier to modify the network as per requirement of the specific application Most importantly, the presented framework is very easy to install, so it can be recommended for real world applications due to its ease of use.

Since in dance forms facial expression plays an important role, there is ample scope of applying image processing based gesture recognition strategy that would take care of the facial expression as well.

## 6.7   Matlab Codes

```
% Code is written by Rimita Lahiri
% Under the guidance of Sriparna Saha and Prof. Amit Konar
dims=3;
pent_features=5;
joints=20;
fid = fopen('anger0.txt');
A = fscanf(fid, '%g', [1 inf]);
```

```
P=1;
total=length(A);
frames=total/(dims*joints);
F=zeros(frames,joints,dims);
for i=1:frames
    for j=1:joints
        for k=1:dims
            F(i,j,k)=A(P);
            P=P+1;
        end
    end
end
fclose(fid);
%%
i=50; %<<<--------------frame no.
P1=[F(i,1,1),F(i,1,2),F(i,1,3)];
P2=[F(i,2,1),F(i,2,2),F(i,2,3)];
P3=[F(i,3,1),F(i,3,2),F(i,3,3)];
n_ref=cross(P1-P2, P1-P3);
d_ref=P1(1)*n_ref(1)+P1(2)*n_ref(2)+P1(3)*n_ref(3);
P1=[F(i,6,1),F(i,6,2),F(i,6,3)];
P2=[F(i,7,1),F(i,7,2),F(i,7,3)];
P3=[F(i,8,1),F(i,8,2),F(i,8,3)];
n_LA=cross(P1-P2, P1-P3);
d_LA=P1(1)*n_LA(1)+P1(2)*n_LA(2)+P1(3)*n_LA(3);
P1=[F(i,10,1),F(i,10,2),F(i,10,3)];
P2=[F(i,11,1),F(i,11,2),F(i,11,3)];
P3=[F(i,12,1),F(i,12,2),F(i,12,3)];
n_RA=cross(P1-P2, P1-P3);
d_RA=P1(1)*n_RA(1)+P1(2)*n_RA(2)+P1(3)*n_RA(3);
P1=[F(i,14,1),F(i,14,2),F(i,14,3)];
P2=[F(i,15,1),F(i,15,2),F(i,15,3)];
P3=[F(i,16,1),F(i,16,2),F(i,16,3)];
n_LL=cross(P1-P2, P1-P3);
d_LL=P1(1)*n_LL(1)+P1(2)*n_LL(2)+P1(3)*n_LL(3);
P1=[F(i,18,1),F(i,18,2),F(i,18,3)];
P2=[F(i,19,1),F(i,19,2),F(i,19,3)];
P3=[F(i,20,1),F(i,20,2),F(i,20,3)];
n_RL=cross(P1-P2, P1-P3);
d_RL=P1(1)*n_RL(1)+P1(2)*n_RL(2)+P1(3)*n_RL(3);
x=-100:100;
y=-100:100;
[X,Y]=meshgrid(x,y);
```

```
Z1=(d_ref-(n_ref(1)*X)-(n_ref(2)*Y))/n_ref(3);
Z2=(d_LA-(n_LA(1)*X)-(n_LA(2)*Y))/n_LA(3);
Z3=(d_RA-(n_RA(1)*X)-(n_RA(2)*Y))/n_RA(3);
Z4=(d_LL-(n_LL(1)*X)-(n_LL(2)*Y))/n_LL(3);
Z5=(d_RL-(n_RL(1)*X)-(n_RL(2)*Y))/n_RL(3);
figure,mesh(X,Y,Z1)
hold on
mesh(X,Y,Z2)
mesh(X,Y,Z3)
mesh(X,Y,Z4)
mesh(X,Y,Z5)
colormap hsv
colorbar
```

**Results**: Sample run results are given in Fig. 5.6.

```
feature=[];
for i=1:frames
    P1=[F(i,1,1),F(i,1,2),F(i,1,3)];
    P2=[F(i,2,1),F(i,2,2),F(i,2,3)];
    P3=[F(i,3,1),F(i,3,2),F(i,3,3)];
    n_ref=cross(P1-P2, P1-P3);
    d_ref=P1(1)*n_ref(1)+P1(2)*n_ref(2)+P1(3)*n_ref(3);
    P1=[F(i,6,1),F(i,6,2),F(i,6,3)];
    P2=[F(i,7,1),F(i,7,2),F(i,7,3)];
    P3=[F(i,8,1),F(i,8,2),F(i,8,3)];
    n_LA=cross(P1-P2, P1-P3);
    d_LA=P1(1)*n_LA(1)+P1(2)*n_LA(2)+P1(3)*n_LA(3);
    P1=[F(i,10,1),F(i,10,2),F(i,10,3)];
    P2=[F(i,11,1),F(i,11,2),F(i,11,3)];
    P3=[F(i,12,1),F(i,12,2),F(i,12,3)];
    n_RA=cross(P1-P2, P1-P3);
    d_RA=P1(1)*n_RA(1)+P1(2)*n_RA(2)+P1(3)*n_RA(3);
    P1=[F(i,14,1),F(i,14,2),F(i,14,3)];
    P2=[F(i,15,1),F(i,15,2),F(i,15,3)];
    P3=[F(i,16,1),F(i,16,2),F(i,16,3)];
    n_LL=cross(P1-P2, P1-P3);
    d_LL=P1(1)*n_LL(1)+P1(2)*n_LL(2)+P1(3)*n_LL(3);
    P1=[F(i,18,1),F(i,18,2),F(i,18,3)];
    P2=[F(i,19,1),F(i,19,2),F(i,19,3)];
    P3=[F(i,20,1),F(i,20,2),F(i,20,3)];
    n_RL=cross(P1-P2, P1-P3);
    d_RL=P1(1)*n_RL(1)+P1(2)*n_RL(2)+P1(3)*n_RL(3);
```

```
a=((n_ref(1))*(n_LA(1)))+((n_ref(2)*n_LA(2)))+((n_ref(3)*n_LA(3)));
b=sqrt(((n_ref(1))^2)+((n_ref(2))^2)+((n_ref(3))^2));
    c=sqrt(((n_LA(1))^2)+((n_LA(2))^2)+((n_LA(3))^2));
    angle1=(cos((abs(a))/(b*c)))*180/pi;
%angle bet ref and LA planes
a=((n_ref(1))*(n_RA(1)))+((n_ref(2)*n_RA(2)))+((n_ref(3)*n_RA(3)));
b=sqrt(((n_ref(1))^2)+((n_ref(2))^2)+((n_ref(3))^2));
    c=sqrt(((n_RA(1))^2)+((n_RA(2))^2)+((n_RA(3))^2));
    angle2=(acos((abs(a))/(b*c)))*180/pi;
%angle bet ref and RA planes
a=((n_ref(1))*(n_LL(1)))+((n_ref(2)*n_LL(2)))+((n_ref(3)*n_LL(3)));
b=sqrt(((n_ref(1))^2)+((n_ref(2))^2)+((n_ref(3))^2));
    c=sqrt(((n_LL(1))^2)+((n_LL(2))^2)+((n_LL(3))^2));
    angle3=(acos((abs(a))/(b*c)))*180/pi;
%angle bet ref and LL planes
a=((n_ref(1))*(n_RL(1)))+((n_ref(2)*n_RL(2)))+((n_ref(3)*n_RL(3)));
b=sqrt(((n_ref(1))^2)+((n_ref(2))^2)+((n_ref(3))^2));
    c=sqrt(((n_RL(1))^2)+((n_RL(2))^2)+((n_RL(3))^2));
    angle4=(acos((abs(a))/(b*c)))*180/pi;
%angle bet ref and RL planes
    p=[F(i,2,1),F(i,2,2),F(i,2,3)];
a=((n_LA(1))*(p(1)))+((n_LA(2))*(p(2)))+((n_LA(3))*(p(3)))-d_LA;
    b=sqrt(((n_LA(1))^2)+((n_LA(1))^2)+((n_LA(1))^2));
    dist1=abs(a/b);
%normal distance bet spine and LA plane
    a=((n_RA(1))*(p(1)))+((n_RA(2))*(p(2)))+((n_RA(3))*(p(3)))-d_RA;
    b=sqrt(((n_RA(1))^2)+((n_RA(1))^2)+((n_RA(1))^2));
    dist2=abs(a/b);
%normal distance bet spine and RA plane
a=((n_LL(1))*(p(1)))+((n_LL(2))*(p(2)))+((n_LL(3))*(p(3)))-d_LL;
    b=sqrt(((n_LL(1))^2)+((n_LL(1))^2)+((n_LL(1))^2));
    dist3=abs(a/b);
%normal distance bet spine and LL plane
a=((n_RL(1))*(p(1)))+((n_RL(2))*(p(2)))+((n_RL(3))*(p(3)))-d_RL;
    b=sqrt(((n_RL(1))^2)+((n_RL(1))^2)+((n_RL(1))^2));
    dist4=abs(a/b);
%normal distance bet spine and RL plane
feature=[feature,angle1,angle2,angle3,angle4,dist1,dist2,dist3,dist4];
end
```

**Results**: Sample run results are given in Eq. 6.6.

```
tic
%dataset for training creation%
t1=rand(8,120);
t2=rand(8,120);
t3=rand(8,120);
t4=rand(8,120);
t5=rand(8,120);
training1=horzcat(t1,t2,t3,t4,t5);
%test data unknown%
t6=rand(8,120);
%target class creation for testing%
a1=ones(1,120);
b1=zeros(1,480);
tr1=horzcat(a1,b1);
b1=zeros(1,360);
b2=zeros(1,120);
tr2=horzcat(b2,a1,b1);
b1=zeros(1,240);
b2=zeros(1,240);
tr3=horzcat(b2,a1,b1);
b1=zeros(1,360);
b2=zeros(1,120);
tr4=horzcat(b1,a1,b2);
b1=zeros(1,480);
tr5=horzcat(b1,a1);
target1=vertcat(tr1,tr2,tr3,tr4,tr5);
%extra target creation for checking%
target2=horzcat((1*ones(1,120)),(2*ones(1,120)),(3*ones(1,120)),
(4*ones(1,120)),(5*ones(1,120)));
tar=ind2vec(target2);
%network creation and training%
net=newrbe(training1,tar,1.5);
view(net);
%networktesting%
y=sim(net,t6);
y1=vec2ind(y);
%plotting
figure,plotconfusion(target2(:,1:120),y1);
toc
```

**Results**: Sample run results are given in Fig. 6.6 and Table 6.4.

# References

1. D.S. Alexiadis, P. Kelly, P. Daras, N.E. O'Connor, T. Boubekeur, M. Ben Moussa, Evaluating a dancer's performance using kinect-based skeleton tracking, in *Proceedings of the 19th ACM international conference on Multimedia*, 2011, pp. 659–662

2. S. Saha, S. Ghosh, A. Konar, A.K. Nagar, Gesture recognition from indian classical dance using kinect sensor, in *Fifth International Conference on Computational Intelligence, Communication Systems and Networks (CICSyN)*, *2013*, pp. 3–8

3. S.G. Wu, F.S. Bao, E. Y. Xu, Y.-X. Wang, Y.-F. Chang, Q.-L. Xiang, A leaf recognition algorithm for plant classification using probabilistic neural network, in *IEEE International Symposium on Signal Processing and Information Technology*, 2007, pp. 11–16

4. L. Shang, D.-S. Huang, J.-X. Du, C.-H. Zheng, Palmprint recognition using fast ICA algorithm and radial basis probabilistic neural network. Neurocomputing **69**(13), 1782–1786 (2006)

5. D.F. Specht, Probabilistic neural networks for classification, mapping, or associative memory, in *IEEE International Conference on Neural Networks,* 1988, pp. 525–532

6. D.F. Specht, Probabilistic neural networks. Neural Netw. **3**(1), 109–118 (1990)

7. S. Waldherr, R. Romero, S. Thrun, A gesture based interface for human-robot interaction. Auton. Robots **9**(2), 151–173 (2000)

8. S. Essid, X. Lin, M. Gowing, G. Kordelas, A. Aksay, P. Kelly, T. Fillon, Q. Zhang, A. Dielmann, V. Kitanovski, A multi-modal dance corpus for research into interaction between humans in virtual environments. J. Multimodal User Interfaces **7**(1–2), 157–170 (2013)

9. D. Xu, A neural network approach for hand gesture recognition in virtual reality driving training system of SPG, in *18th International Conference on Pattern Recognition, ICPR 2006*, 2006, vol. 3, pp. 519–522

10. K. Murakami, H. Taguchi, Gesture recognition using recurrent neural networks, in *Proceedings of the SIGCHI conference on Human factors in computing systems*, 1991, pp. 237–242

11. K.Z. Mao, K.-C. Tan, W. Ser, Probabilistic neural-network structure determination for pattern classification. Neural Netw. IEEE Trans. **11**(4), 1009–1016 (2000)

12. A. LaViers, Y. Chen, C. Belta, M. Egerstedt, Automatic sequencing of ballet poses. Robot. Autom. Mag. IEEE **18**(3), 87–95 (2011)

13. F. Guo, G. Qian, Dance posture recognition using wide-baseline orthogonal stereo cameras, in *7th International Conference on Automatic Face and Gesture Recognition, FGR 2006*, 2006, pp. 481–486

14. B. Peng, G. Qian, S. Rajko, View-invariant full-body gesture recognition via multilinear analysis of voxel data, in *Third ACM/IEEE International Conference on Distributed Smart Cameras, ICDSC 2009*, 2009, pp. 1–8

15. Y. Lee, K. Jung, Non-temporal mutliple silhouettes in hidden Markov model for view independent posture recognition, in *International Conference on Computer Engineering and Technology, ICCET'09*. 2009, vol. 1, pp. 466–470

16. P. Silapasuphakornwong, S. Phimoltares, C. Lursinsap, A. Hansuebsai, "Posture recognition invariant to background, cloth textures, body size, and camera distance using morphological geometry," in *International Conference on Machine Learning and Cybernetics (ICMLC)*, 2010, vol. 3, pp. 1130–1135

17. F. Buccolieri, C. Distante, A. Leone, Human posture recognition using active contours and radial basis function neural network, in *IEEE Conference on Advanced Video and Signal Based Surveillance, AVSS 2005*, 2005, pp. 213–218

18. N.M. Tahir, A. Hussain, S.A. Samad, H. Hussin, On the use of decision tree for posture recognition, in *International Conference on Intelligent Systems, Modelling and Simulation (ISMS)*, 2010, pp. 209–214

19. A. Konar, *Artificial Intelligence and Soft Computing: Behavioral and Cognitive Modeling of the Human Brain* (vol. 1. CRC Press, Boca Raton, 1999)
20. A. Alvarez-Alvarez, G. Trivino, O. Cordón, Body posture recognition by means of a genetic fuzzy finite state machine, in *IEEE 5th International Workshop on Genetic and Evolutionary Fuzzy Systems (GEFS)*, 2011, pp. 60–65
21. K.K. Htike, O.O. Khalifa, Comparison of supervised and unsupervised learning classifiers for human posture recognition, in *International Conference on Computer and Communication Engineering (ICCCE)*, 2010, pp. 1–6
22. T. Shiratori, A. Nakazawa, K. Ikeuchi, Synthesizing dance performance using musical and motion features, in *Proceedings 2006 IEEE International Conference on Robotics and Automation, ICRA 2006*. 2006, pp. 3654–3659
23. L. Kovar, M. Gleicher, F. Pighin, Motion graphs, in *ACM SIGGRAPH 2008 classes*, 2008, p. 51
24. K. Pullen, C. Bregler, Motion capture assisted animation: Texturing and synthesis. ACM Trans. Graph. (TOG) **21**(3), 501–508 (2002)
25. O. Arikan, D.A. Forsyth, Interactive motion generation from examples. ACM Trans. Graph. (TOG) **21**(3), 483–490 (2002)
26. K. Grochow, S.L. Martin, A. Hertzmann, Z. Popović, Style-based inverse kinematics. ACM Trans. Graph. (TOG) **23**(3), 522–531 (2004)
27. T. Kim, S. Il Park, S.Y. Shin, Rhythmic-motion synthesis based on motion-beat analysis. ACM Trans. Graph. (TOG) **22**(3), pp. 392–401 (2003)
28. M. Stone, D. DeCarlo, I. Oh, C. Rodriguez, A. Stere, A. Lees, C. Bregler, Speaking with hands: Creating animated conversational characters from recordings of human performance. ACM Trans. Graph. **23**(3), 506–513 (2004)
29. M.-H. Lee, U.-M. Kim, J.-I. Park, An analysis methodology on emotion of Korean traditional dance using a virtual reality system, in *Second International Conference on Culture and Computing (Culture Computing)*, 2011, pp. 149–150
30. J.A. Russell, L.F. Barrett, Core affect, prototypical emotional episodes, and other things called emotion: dissecting the elephant. J. Pers. Soc. Psychol. **76**(5), 805 (1999)
31. J. Posner, J.A. Russell, B.S. Peterson, The circumplex model of affect: an integrative approach to affective neuroscience, cognitive development, and psychopathology. Dev. Psychopathol. **17**(3), 715–734 (2005)
32. J.A. Russell, M. Lewicka, T. Niit, A cross-cultural study of a circumplex model of affect. J. Pers. Soc. Psychol. **57**(5), 848 (1989)
33. S. Saha, M. Pal, A. Konar, R. Janarthanan, Neural network based gesture recognition for elderly health care using kinect sensor, in *Swarm, Evolutionary, and Memetic Computing*, Springer, 2013, pp. 376–386
34. S. Saha, S. Datta, A. Konar, R. Janarthanan, A study on emotion recognition from body gestures using Kinect sensor, in *International Conference on Communications and Signal Processing (ICCSP)*, 2014, pp. 56–60

# Chapter 7
# Differential Evolution Based Dance Composition

In this chapter, we propose a novel approach in which a system autonomously composes dance sequences from previously taught dance moves with the help of the well-known differential evolution algorithm. In this chapter, we propose a novel approach in which a system autonomously composes dance sequences from previously taught dance moves with the help of the well-known differential evolution algorithm. Initially, we generated a large population of dance sequences. The fitness of each of these sequences was determined by calculating the total inter-move transition abruptness of the adjacent dance moves. The transition abruptness was calculated as the difference of corresponding slopes formed by connected body joint co-ordinates. By visually evaluating the dance sequences created, it was observed that the fittest dance sequence had the least abrupt inter-move transitions. Computer simulation undertaken revealed that the developed dance video frames do not have significant inter-move transition abruptness between two successive frames, indicating the efficacy of the proposed approach. Gestural data specific of dance moves is captured using a Microsoft Kinect sensor. The algorithm developed by us was used to fuse the dancing styles of various 'Odissi' dancers dancing to the same rasa (theme) and tala (beats) and loy (rhythm). In future, it may be used to fuse different forms of dance.

## 7.1 Introduction

Gestures are used as alternative form of expression to enrich or augment vocal and facial expressions. As gestures are a part of natural communication, gestural interfaces may be viewed as the pre-requisites of ambient computing. Dance is a special form of gestural expression portrayed in a rhythmic form to generally

---

Contributed by Reshma Kar, Amit Konar and Aruna Chakraborty

© Springer International Publishing AG 2018
A. Konar and S. Saha, *Gesture Recognition*, Studies in Computational
Intelligence 724, DOI 10.1007/978-3-319-62212-5_7

communicate the context of a music piece. An important part of dance is establishing a smooth flow of dance moves to create a visually appealing sequence of movements.

Dance is specific of different regions, communities and personal styles. Dance schools world-wide teach different dance moves to students, which they are often required to combine to create new dance sequences. It is an important part of dance creation for teachers and self-learning for students. Various techniques have been proposed to ease dance choreography and most of these works have been tested on the dance form 'Ballet' [1, 2]. However, not much work has been done on composing Indian dance forms. In this work, we focus our attention on an Indian classical dance form known as 'Odissi' which is a classical dance of the state Orissa.

Choreographing dance sequences is definitely an art which different people succeed to different degrees. Thus, it would be an interesting task to see how high-end computational algorithms perform at choreographing dance sequences. Naturally, in a visually appealing and easily executable dance sequence, the steps following each other would execute smooth transitions. In our work, we used the same constraint to judge randomly generated dance sequences and finally select the best among them.

At its heart, computational creativity is the study of building software that exhibits behaviour that would be deemed creative in humans [3]. Widmer et al. [4] described a computer which can play a musical piece expressively by shaping the musical parameters like tempo and dynamics. In [5], Gervas provided an excellent review of different computational approaches to storytelling. We can find a brief glimpse of the computational attempts at music composition in [6]. Similarly, computational dance composition techniques find mention in the works of [7, 8]. Some works on ballet are discussed in the following paragraphs however their approach is different from ours as they compose dances from pre-defined choreography structures characteristic of 'ballet'. In contrast, dance composition in Indian classical dance forms are more dependent on the context of the music and follow much less rigid sequencing.

A 3-dimensional animation system was developed based on hierarchic human body techniques and its motion data in [9]. Using the proposed system, One can easily compose and simulate classic ballet dance in real-time on the internet. Their goal was to develop and integrate several modules into a system capable of animating realistic virtual ballet stages in a real-time performance. This includes modeling and representing virtual dancers with high realism, as well as simulating choreography and various stage effects. For handling motion data more easily, they developed a new action description code based on "Pas" which is a set of fundamental movements for classic ballet.

An automatic composition system for ballet choreographies was developed by using 3DCG animation in [1]. Their goal was to develop some useful tools in dance education such as creation-support system for ballet teachers and self-study system for students. The algorithm for automatic composition was integrated to create utilitarian choreographies. As a result of an evaluation test, they verified that the created choreographies had a possibility to be used in the actual lessons. This system is valuable for online virtual dance experimentation and exploration by teachers and choreographers involved in creative practices, improvisation, creative movement, or dance composition. The goals, technical novelty and claims of the chapter are briefly given below.

*Goals:*

- *To choreograph dance sequence from previously taught dance sequences.*
- *To study the applicability of a heuristic algorithm in composing dance.*
- *To be able to automatically identify visually appealing dance sequences.*

*Technical Novelty:*

- *A novel measure of fitness of dance sequences is presented.*
- *Differential evolution is employed to solve the problem of creating a smooth flowing pattern of dance.*
- *A novel metric of comparison of different forms of dance is introduced.*

*Claims:*

- *Heuristic Algorithms can be used to choreograph dance.*
- *A visually appealing and easy to execute dance sequence can be recognized using the proposed metric for measuring inter gesture transition abruptness.*
- *Dances of different forms can be quantitatively compared on a common platform using a metric which evaluates the dances forms based on their commonalities.*

## 7.2   Differential Evolution Algorithm

Proposed by Storn and Prince [10–12] in 1995, Differential Evolution (DE) Algorithm was found to outperform many of its contemporary heuristic algorithms [13]. Following its discovery, the structural simplicity and efficiency of the algorithm attracted researchers, for use in optimization of rough and multi-modal objective functions. Several extensions of the basic DE algorithm are reported in the literature. The De/rand/1 version of the DE algorithm, containing four main steps are outlined below.

### 7.2.1  Initialization

DE starts with *NP* number of *D*-dimensional parameter vectors, selected in a uniformly random manner from a prescribed search space, given for an engineering optimization problem. The parameter vectors here represent trial solutions of optimization. The *i*th parameter vector of the population at the current generation *G* is formally represented by

$$\vec{Z}_i(G) = \left[ z_{i,1}(G), z_{i,2}(G), z_{i,3}(G), \ldots, z_{i,D}(G) \right] \tag{7.1}$$

where each component of $\vec{Z}_i(G)$ lies between corresponding components of $\vec{Z}_{min}$ and $\vec{Z}_{max}$ given by $\vec{Z}_{min} = [z_{min-1}, z_{min-2}, z_{min-3}, \ldots, z_{min-D}]$ and $\vec{Z}_{max} = [z_{max-1}, z_{max-2}, z_{max-3}, \ldots, z_{max-D}]$.

The *j*th component of the *i*th vector is given by

$$z_{i,j}(0) = z_{j-min} + rand_{i,j}(0,1) \times (z_{j-max} - z_{j-min}) \tag{7.2}$$

where $rand_{i,j}(0,1)$ is a uniformly distributed random number lying between 0 and 1 and is instantiated independently for each component of the *i*th parameter vector.

### 7.2.2  Mutation

For each individual population target vector $\bar{z}_i(G)$ belonging to population *G*, a set of three other randomly selected vectors: $\vec{Z}_j(G), \vec{Z}_k(G)$ and $\vec{Z}_l(G))$ are chosen and arithmetically recombined to create a mutant vector.

$$\vec{V}_i(G) = \vec{Z}_j(G) + F(\vec{Z}_k(G) - \vec{Z}_l(G)) \tag{7.3}$$

The process is repeated for each parameter vector i in [1, *NP*].

### 7.2.3  Recombination

Recombination, allows each pair of trial vector $\vec{Z}_i(G)$ to recombine with its corresponding mutant vector $\vec{V}_i(G)$ for i = 1 to NP. Here, recombination is performed position-wise over the length of the parameter vectors. Thus, for trial vectors of length NP, we need to perform NP number of recombination. The principle of recombination for a single position j is given below. Let $rand_{i,j}$ be a uniformly distributed random number in [0, 1] used to select the right element from $V_{i,j}(G)$ and $Z_{i,j}(G)$ into $j$th position of the Target vector $u_{i,j}(G)$, where

$$u_{i,j}(G) = \begin{cases} v_{i,j}(G) & if\ rand_{i,j} \leq Cr\ or\ j = j_{rand} \\ z_{i,j}(G) & otherwise \end{cases} \tag{7.4}$$

and $Cr$ denotes the crossover rate, which is defined at the beginning of the program once only with a typical value of 0.7.

### 7.2.4 Selection

In this step, survival of the fittest chromosome is ensured where the offspring replaces the parent vector, if it yields a better value for the objective function.

$$\begin{aligned} \vec{Z}_i(G+1) &= \vec{U}_i(G) && if\ f(\vec{U}_i(G)) \leq f(\vec{Z}_i(G)) \\ &= \vec{Z}_i(G) && if\ f(\vec{U}_i(G)) > f(\vec{Z}_i(G)) \end{aligned} \tag{7.5}$$

Repeat from step 2 until stopping criterion, defined on convergence on the best fit member of the trial solutions or convergence of the average fitness over iterations or a fixed large number of iterations are met.

## 7.3 Metric to Compare Dance Gestures

Our basic principle in placing dance-gestures adjacently in visually appealing dance choreography is based on a simple comparison metric of these gestures. The dance gestures are video clips of variable duration ranging from 6 to 8 s. Since the Kinect records at the speed of approximately 30 frames per second, the number of frames captured for each dance-gesture is different. Each video clip, comprising a number of frames, is decoded into a sequence of skeletal diagrams, where each skeletal diagram corresponds to an individual frame (illustrated in Fig. 7.1). These skeletal diagrams are sketches of the body structure indicated by 3D (three dimensional) straight line segments joining 20 fundamental junctions of the dancer's physique (Fig. 7.2). Each 3D straight line segment is projected on to XY, YZ and ZX planes

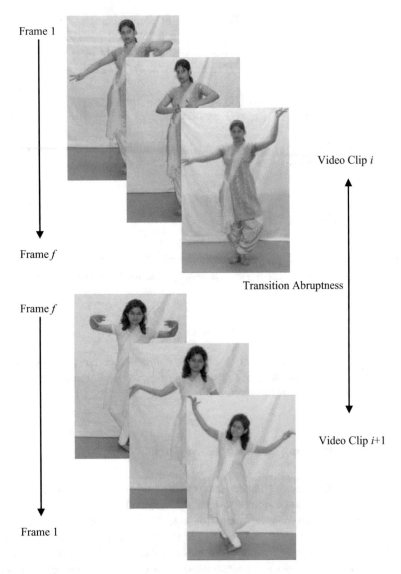

**Fig. 7.1** Video clips of dance gestures for different frames

as illustrated in Fig. 7.3a–c, and the slopes of the projected 2D straight lines in the three planes with respect to X-, Y- and Z-axes respectively are computed. These slopes are used as metrics to compare two different 3D straight lines representing the orientation of one given body part (say, the right forearm) present in two frames of two videos. While constructing a new dance video from the existing video clips

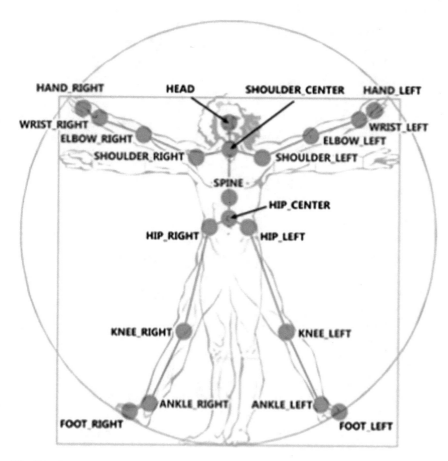

**Fig. 7.2** Twenty body joint co-ordinates obtained from Kinect sensor

of shorter durations, we need to match the last frame and the first frame of two consecutive videos, which might come in order to ultimately offer a new dance video. Naturally, the matching of 3D straight lines is required between the last frame and the first frame of each pair of videos to test their possible just-apposition in the final video.

We introduced a new term called, Transition Abruptness (TA), to measure the abruptness of inter-gesture transitions in a dance sequence. It is a measure of the total difference between two 3-dimensional body skeletal structures, induced by comparing the angular difference of the projected straight lines of each 3D straight line link of two skeleton structures. TA between two dynamic gestures, $G_i$ and $G_{i+1}$,

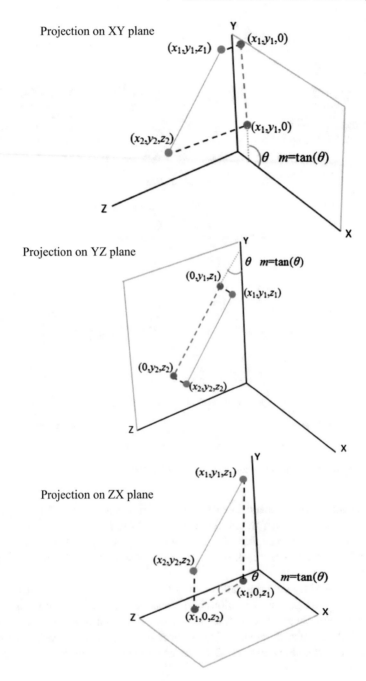

**Fig. 7.3** Calculation of slope of projections of a line

each comprising a number of sequential frames, is measured by (7.6), where $p_{x,i,j,k}$ denotes the slope of the projected straight line $k$ on the X-Y plane with respect to x-axis of the $j$th frame of the $i$th gesture. The parameters: $q_{y,i,j,k}$ (and $r_{z,i,j,k}$) denote slopes of the projected straight line $k$ on the Y-Z plane (Z-X plane) with respect to y-axis (z-axis) of the $j$th frame of the $i$th gesture.

$$TA(G_i, G_{i+1}) = \sum_{\forall k} \sum_{\forall i, \forall j} (|p_{x,i,1,k} - p_{x,i+1,f,k}|$$

$$+ |q_{y,i,1,k} - q_{y,i+1,f,k}| + |r_{z,i,1,k} - r_{z,i+1,f,k}|) \tag{7.6}$$

The *Total Transition Abruptness* of the gesture permutation $G_1, G_2, G_3 \ldots G_n$ is calculated in (7.7) by summing TA of each two consecutive dance gestures

$$\text{Transition Total Abruptness} = \sum_{\forall i} TA(G_i, G_{i+1}) \tag{7.7}$$

## 7.4 Selection of Dance Sequences

Evolutionary Algorithms are generally employed to optimize complex non-linear, multi-dimensional objective functions which usually contain multiple local optima. In our experiments, the search surface considered is characteristic of dance-move permutations. The fitness function which ensures the survival of best dance sequences is obtained based on the total abruptness in inter-gesture transitions. In this section, we present a novel scheme of dance composition by selection of dance permutations. The basic structure of parameter vectors used here is of 3 K dimension, where K denotes the maximum value of k. The integer 3 appears due to inclusion of three elements: $p_{x,i,j,k}$, $q_{y,i,j,k}$ and $r_{z,i,j,k}$ each $K$-times in the parameter vector. It is important to note that we have two sets of parameter vectors, the first set for representation of the first frame of each video clip describing a dynamic gesture, and the last set to represent the last frame of the same dynamic gesture. In the evolutionary algorithm we match two dynamic gestures by measuring the total transition abruptness of the last frame and the first frame of any two video clips. The pseudo code of the proposed gesture selection algorithm is outlined next.

**Pseudo Code:**
**Input:** A set of body joint co-ordinates of $n$ dance gestures $G_1, G_2 \ldots . G_5$.
**Output:** A visually appealing permutation of $m$ dance gestures.
**Begin**
    1.     Randomly initialize a population of NP solutions (dance sequences)

    $\vec{Z}_j(g) = \{G_1, G_2 \ldots, G_5\}$. Here j runs from 1 to NP and g signifies generation

    2.     **For** Iter=1 to MaxIter
         **Begin**
             a)     Create mutant vector using mutation scheme section 2.
             b)     Crossover target vector and mutant vector as outlined in section 2.
             c)     Evaluate Fitness by equation 7 i.e.

$$fitness = \sum_{i=1}^{n} Transition\ Abryptness(\vec{G}_i, \vec{G}_{i+1}) \quad \forall G_i \in \vec{Z}$$

             d)   **If** trial vector is better (less abrupt) than target
             **Then** replace trial by target in the next generation
             **End If**
         **End For**
    3.     Return $\vec{Z}$ after minima value of fitness function converges.
**End**
**_Procedure:_** Transition Abruptness ( $\vec{G}_i, \vec{G}_{i+1}$ )

**Input:** $\vec{G}_i, \vec{G}_{i+1}$

**Output:** Inter-gesture Transition Abruptness TA
**Begin**
Let K be the number of lines in each skeletal diagram
    1.    **For** k=1 to K
             a)    Extract $l^k_{G_i}, l^k_{G_i+1}$ where $l^k_{G_i} \in last\ frame\ of(\vec{G}_i)$

                and

             $l^k_{G_i+1} \in first\ frame\ of(\vec{G}_i+1)$

             Here each of the above lines are represented by ,the projections of
             their slopes ($p_{x,i,j,k}$, $q_{y,i,j,k}$, $r_{z,i,j,k}$ discussed preceding eq. 6)

             b)    Compute the fitness of each consecutive pair of gestures by Eq.6
                 $A_l = TA(G_i, G_{i+1})$
**End For**
    2.   TA=0

    **For** l=1 to N-1
             a)    $TA = TA + A_l$
**End For**
    3.   **Return** TA

## 7.5  Comparison Metric of Different Dance Forms

The comparison of dance composition algorithms is a difficult task as there is no numerical formula to distinguish the quality of one dance from another. To the best of our knowledge, ours is the first work on *Odissi* dance composition which makes the comparison of our algorithm further difficult. Thus, we also propose a metric for comparison of dance of different forms and use it to compare dance composed by our algorithm with dance composed by other algorithms. Apparently, a good dance composition technique maintains a smooth flow of transitions between gestures and also a high amount of dynamism. The smoothness of inter-gesture transitions can be measured by the calculating *Transition Abruptness* between two consecutive gestures. A smaller *Transition Abruptness* between consecutive gestures indicates a smooth flow of dance. Similarly, if non-adjacent gestures have higher value of *Transition Abruptness* the dance can be said to have more dynamic steps. Thus, we define the *Visual Appeal* of a dance sequence $S$ as follows composed of $n$ gestures $G_i$, where $i$ ranges from 1 to $n$.

$$Visual \ Appeal(S) = \frac{\sum_{i=1}^{n} Transition \ Abruptness \ (G_i, G_{i+2})}{\sum_{i=1}^{n} Transition \ Abruptness \ (G_i, G_{i+1})} \qquad (7.8)$$

## 7.6  Experiments

The Microsoft's Kinect captures skeletal co-ordinates at an approximate rate of 30 frames per second, with the help of an RGB camera, an infrared projector a monochrome CMOS (complimentary metal-oxide semiconductor) sensor. This configuration allows the *Kinect* to efficiently capture the 3-dimensional skeletal co-ordinates in closed room settings. However, certain precautions are required to be maintained to ensure noise-free collection of data. A white background is maintained to eliminate possible chances of interference of background objects in data collection. To ensure good quality of images, two standing lights are placed facing the dance platform. The dancers are also instructed to avoid wearing clothes which are too loose around the body joints. The subjects should be within an appropriate range of the *Kinect* which is approximately 1.2–3.5 m or 3.9–11 ft.

In our experiments, 10 experienced *Odissi* dancers are asked to perform 5 of their preferred dance gestures which are each captured separately with the help of the Microsoft *Kinect*. Thus, we have a total of 50 video clips of dynamic dance gestures. A sequential permutation of 5 of these dance gestures are then selected from a large amount of such randomly created permutations with the help of *differential evolution* algorithm. It is implicit that dancing style of each dancer is

**Fig. 7.4** Gesture at different instances for an Odissi dancer

dependent on various factors including personal choice and interpretation of dance steps. This lends a unique touch to each of the dancer's gestures though all of them belong to the same dance form. The aim of our experiments is to combine the dance styles of different teachers which has the potential to give rise to interesting patterns and also has future applications in fusing various dance forms. Stills from dance gestures of the different teachers are shown in Fig. 7.4.

The data captured using Kinect represents the bodily movements over time with an approximate distance of 10 cm from ground truth. However, some important considerations must be made before performing analysis of the skeletal co-ordinates obtained. For example two subjects (dancers) may be portraying the same gesture and yet have different X, Y and Z co-ordinates. The main reasons of this are difference in skeletal structure of subjects and variable distance of subjects from Kinect camera. To overcome these differences, we have used the slope of the lines in skeletal diagram as the comparison parameter.

In the next phase, differential evolution algorithm is used to evaluate different permutations of the dance sequences and find the best among them. The comparison

**Table 7.1** Analysis of inter-gesture transition abruptness values obtained

| Figure No. | Transition abruptness (TA) | Analysis |
|---|---|---|
| Figure 7.5a | 292.0031 | It can be easily seen that the poses are quite close |
| Figure 7.5b | 350.9825 | TA increases as gradual slope changes occur |
| Figure 7.5c | 955.3705 | Drastic Changes in both upper and lower body increase TA |
| Figure 7.5d | 900.0376 | Comparatively more changes in lower body than upper body cause slight decrease in TA |

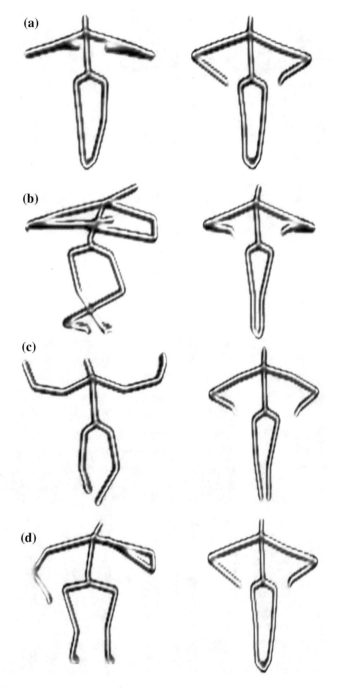

**Fig. 7.5**  Inter-gesture transition poses

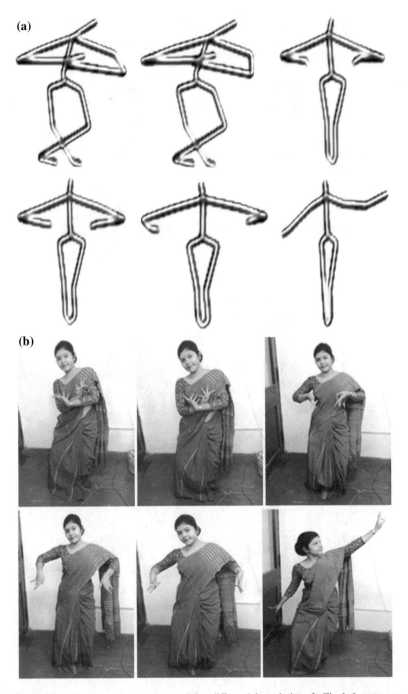

**Fig. 7.6 a** Final dance sequence generated by differential evolution. **b** Final dance sequence generated by differential evolution demonstrated by a dancer's performance

of dance sequences is based on *total transition abruptness*. The total transition abruptness is simply the sum of all *inter-gesture transition abruptness* $TA = (G_i, G_{i+1})$. The comparison for placing the gestures adjacently is based on the first frame of a gesture and the last frame of the preceding gesture. In Fig. 7.5 we have illustrated the diagrams such four pairs; in Table 7.1 we described how the *inter-gesture transition abruptness* (total slope differences of the projections of the indicated skeletal diagrams) are able to detect inter move abruptness of the pairs in Fig. 7.5. In Fig. 7.6 we have shown a part of one of the fittest generated dance sequences.

## 7.7   Performance Evaluation

The convergence of different optimization algorithms on the dance composition problem is illustrated in Fig. 7.7 and the corresponding data is also given in Table 7.2. It is seen that *Differential Evolution* algorithm performs better than 3 other algorithms namely *Adaptive Bee Colony algorithm, Firefly algorithm* and *Particle Swarm Optimization* for the dance composition problem. Using our

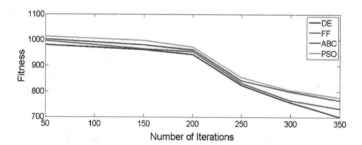

**Fig. 7.7** Value of fitness function of different Heuristic algorithms for dance composition, *DE* Differential Evolution, *FF* Firefly Algorithm, *ABC* Adaptive Bee Colony optimization, *PSO* Particle Swarm Optimization

**Table 7.2** Average fitness value of different heuristic algorithms over different iterations

| Iterations | Fitness | | | |
|---|---|---|---|---|
| | DE | FF | ABC | PSO |
| 50 | 980 | 996 | 1002 | 1013 |
| 150 | 962 | 965 | 980 | 997 |
| 200 | 943 | 954 | 961 | 972 |
| 250 | 824 | 832 | 842 | 855 |
| 300 | 756 | 764 | 799 | 804 |
| 350 | 700 | 732 | 765 | 776 |

**Table 7.3** The comparison of different dance composition algorithms

| Dance composition technique | Average *visual appeal* for 3 runs |
|---|---|
| Ballet composition [16] | 18.67 |
| Contemporary dance composition [1] | 19.99 |
| *Odissi* composition by differential evolution | 23.44 |

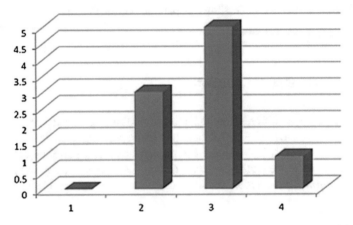

**Fig. 7.8** Performance evaluation of the dance composition system as per the opinion of 10 dancers

proposed metric *visual appeal* outlined in (7.8), we compared our dance composition techniques with two other dance composition techniques for different dance forms [1, 2]. Our algorithm performs substantially better than other algorithms on an average as indicated in Table 7.3. Finally, we asked 10 *Odissi* dancers to rate the dance permutation generated by our system on a scale of 1–4 as follows, 1: Bad, 2: Fair, 3: Good, 4: Excellent. The results obtained are given in Fig. 7.8.

## 7.8 Discussion

This chapter introduced a meta-heuristic approach to compose dance by optimally selecting inter-gestures movement patterns from multiple dance gestures. The fitness estimate used for dance gesture composition ensures a smooth transition of frames in the dance video. One metric representing visual appeal has been introduced to compare the performance of the proposed Differential Evolution (DE) based meta-heuristic algorithm with existing algorithms of dance composition. The proposed DE-based realization is also compared with other standard realizations including, firefly, PSO and ABC with respect to convergence time, and the results indicate that the DE realization outperforms its partners in the present context.

## 7.9  Conclusion

This chapter provides a fresh perspective on the technique of dance composition. Dance composers are commonly required to ensure a smooth flow of inter-gesture dance transitions. Our algorithm uses this common know-how and integrates it into the differential evolution algorithm to compose interesting dance patterns. Since the differential evolution algorithm uses a random initial population each time; the chances of dance patterns generated being novel each time is high. The algorithm can be improved by including other constraints in the fitness function which increase the dynamicity of the dance patterns. There is also a wide scope in creating dance composition with context to music-mood as well as lyrics. Apart from this one may also analyse the relative performance of different evolutionary algorithms in dance composition. In a nutshell, our work area relatively unexplored and offers a wide future scope.

## References

1. A. Soga, B. Umino, M. Hirayama, Automatic composition for contemporary dance using 3D motion clips: Experiment on dance training and system evaluation, in *CW'09. International Conference on CyberWorlds*, 2009, pp. 171–176
2. J. Dancs, R. Sivalingam, G. Somasundaram, V. Morellas, N. Papanikolopoulos, Recognition of ballet micro-movements for use in choreography, in *IEEE/RSJ International Conference on Intelligent Robots and Systems*, 2013, pp. 1162–1167
3. S. Colton, R. de López Mantaras, O. Stock, Computational Creativity: Coming of Age. AI Mag. **30**(3), 11 (2009)
4. G. Widmer, S. Flossmann, M. Grachten, YQX plays Chopin. AI Mag. **30**(3), 35 (2009)
5. P. Gervás, Computational approaches to storytelling and creativity. AI Mag. **30**(3), 49 (2009)
6. M. Edwards, Algorithmic composition: computational thinking in music. Commun. ACM **54**(7), 58–67 (2011)
7. S.F. de Sousa Junior, M.F.M. Campos, Shall we dance? A music-driven approach for mobile robots choreography, in *IEEE/RSJ International Conference on Intelligent Robots and Systems*, 2011, pp. 1974–1979
8. S. Jadhav, M. Joshi, J. Pawar, Art to SMart: an evolutionary computational model for BharataNatyam choreography, in *12th International Conference on Hybrid Intelligent Systems (HIS)*, 2012, pp. 384–389
9. A. Soga, M. Endo, T. Yasuda, Motion description and composing system for classic ballet animation on the Web, in *Proceedings of 10th IEEE International Workshop on Robot and Human Interactive Communication*, 2001, pp. 134–139
10. R. Storn, K. Price, Differential evolution–a simple and efficient heuristic for global optimization over continuous spaces. J. Glob. Optim. **11**(4), 341–359 (1997)
11. R. Storn, K. Price, *Differential Evolution—A Simple and Efficient Adaptive Scheme for Global Optimization Over Continuous Spaces*. ICSI Berkeley, 1995
12. K. Price, R. M. Storn, J. A. Lampinen, *Differential Evolution: A Practical Approach to Global Optimization* (Springer, Berlin, 2006)
13. J. Vesterstrom, R. Thomsen, A comparative study of differential evolution, particle swarm optimization, and evolutionary algorithms on numerical benchmark problems, in *CEC2004. Congress on Evolutionary Computation*, 2004, vol. 2, pp. 1980–1987

# Chapter 8
# EEG-Gesture Based Artificial Limb Movement for Rehabilitative Applications

Due to damage in parietal and/or motor cortex regions of a rehabilitative patient (subject), there can be failure while performing day-to-day basic task. Thus this chapter focuses on directing an artificial limb based on brain signals and body gesture to assist the subject. This research finds tremendous applications in rehabilitative aid for the disable persons. To concretize our goal we have developed an experimental setup, where the target subject is asked to catch a ball while his/her brain (occipital, parietal and motor cortex) signals electroencephalography (EEG) sensor and body gestures using Kinect sensor are simultaneously acquired. These two data are mapped using cascade correlation learning architecture to train Jaco robot arm to move accordingly. When a rehabilitative patient is unable to catch the ball, then in that scenario, the artificial limb is helpful for assisting the patient to catch the ball. The proposed system can be implemented not only in ball catching experiment but also in several application areas where an artificial limb needs to perform a locomotive task based on EEG and body gesture.

## 8.1 Introduction

Gestures constitute an important medium for human beings to interact and communicate with the surroundings. Gestures are expressive body movements, involving the face and hands in majority of the cases. Gesture recognition i.e., interpretation of the different human gestures is a very important aspect of human computer interaction. It enables humans to efficiently communicate with machines using only sign language.

Suppose a subject wants to bring a glass of water from a nearby table to drink, but is suffering from damage in the motor cortex region in the brain, then it is not

Contributed by Sriparna Saha, Amit Konar, Anuradha Saha and Arup Kumar Sadhu

© Springer International Publishing AG 2018
A. Konar and S. Saha, *Gesture Recognition*, Studies in Computational
Intelligence 724, DOI 10.1007/978-3-319-62212-5_8

feasible for the subject to bring the glass using his hands [1]. But in this scenario we can place an electroencephalography (EEG) sensor to record the occipital data of that subject while he is looking at that glass and based on this information employ a robotic arm (artificial limb) to fetch the glass. This is just a simple example of how to control an artificial limb using human brain signals, where the position of the target object (i.e., the glass) is stationary. This motivates us to implement our research work in the scenario where the position of the target with respect to time is in motion. Thus we have outlined the work in such a manner where two subjects are involved, one is our target subject wearing a EEG sensor and concurrently whose motion is captured by a motion sensing device, Kinect and another subject is just throwing a ball towards the first subject. We can use this technique in several other cases which involves artificial limb to do some locomotive works based on brain signal and human gestures. The work finds tremendous importance in rehabilitative application areas where a subject is suffering from difficulties in parietal and occipital damage.

There are some existing literatures which deal with human machine interface (HMI) [2]. These papers show how information from gesture and brain signals can be utilized to instruct an artificial limb in several environments. Maheu et al. [1] states the advantages of Jaco arm to do tasks related to rehabilitation. Bhattacharyya et al. proposed how the efficient movement of robot can be done from the acquired EEG signals [3, 4]. All of these literatures are based on directing the robot using EEG signals. But to perform human like action the robot arm need to be train with body gestures simultaneously. Wang et al. shows how human hand gesture plays an important part for robot control [5]. For this purpose surface electromyography (sEMG) and Kinetic signals are taken form upper limb and classification is done by support vector machine (SVM). Teleportation between double hand movements and double-arm robot manipulator is developed by Du et al. where human motion is captured by Kinect [6]. Multimodal interactions including speech, gesture etc. is carried out by Csapo et al. with Nao humanoid robot for HMI applications [7].

The prediction of hand movements acquired using Kinect sensor from EEG signals obtained from respective parts of the brain is done by Datta et al. [8] using principal component analysis (PCA) for feature selection and regression analysis using back propagation learning architecture (BPNN). But this paper does not train any robot to perform the task in case the subject is unable to do the same. Frisoli et al. [9] designed a system based on human gaze, Kinetic data and EEG signal analysis for neuro-rehabilitation of patients. The outcome of that work is to direct a robotic upper limb exoskeleton based on the acquired data in real-time. In [10], Roy et al. has controlled not only position, but also velocity of an exoskeleton using state feedback PI controller. EEG signals containing the right motor intention, as captured from the human brain, is classified using Quadratic discriminant analysis (QDA) classifier and is fed to the controller to execute accurate movement. Another interesting application of Kinect sensor along with EEG signal is to improve the quality of life of patients suffering from motor/communication impairments by training them through gaming technology [11]. The depth segmentation and RGB data obtained from Kinect sensor as well as EEG features including mean absolute value and variance unbiased estimator are utilized to classify hand grasping during gaming activity. Similar kind of experiment for tracking self-paced hand opening/closing as proposed by has been carried out on stroke patients by using EMG sensors along with EEG and Kinect sensors [12].

Siradjuddin et al. has implemented a system to instruct a PowerCube manipulator with seven degrees of freedom employing Kalman filter to predict the goal state based on three dimensional (3D) data from Kinect sensor [13]. Automatic navigation is proposed in where a new approach is taken to develop the depth map and compared with the depth map obtained from Kinect [14].

To implement this novel system based on EEG-gesture synergism for artificial limb movement we require three devices, namely EEG sensor [2–4, 8–12], Kinect [5–9, 11–14] and Jaco arm [1–4, 9]. For healthy subjects, when a ball is thrown, then they are directed to catch the ball. The EEG signals from occipital, parietal and motor cortex regions are acquired using EEG acquisition device as well as the skeleton of the target subject is obtained using Kinect sensor while catching the ball. Now three mappings are done between occipital-parietal, parietal-motor cortex and motor cortex-skeleton using cascade correlation learning architecture (CCLA) [15, 16] in the first training phase. The order of the mappings are in that fashion because first the subject sees the ball coming towards him/her using occipital region, then the parietal region plans how to catch the ball and based on the planning the motor cortex region instructs the hand of the subjects accordingly. Upon getting these signals, the subject tries to catch the ball. For the next training phase, CCLA is employed to map skeleton and displacements of the Jaco arm, such that in the absence of the parietal and motor cortex signals, the artificial limb is able to catch the ball. We have used CCLA as it has much better performance in terms of weight adaptation and randomness with comparison with BPNN. After the completion of the training phase, we have tested our proposed system for rehabilitative patients with damage in parietal and/or motor cortex regions. These subjects due to disability are unable to plan and execute the movements of the hands, thus to provide assistance, Jaco robot arm is orchestrated to catch the ball. Here we have created a bypass network using the data from occipital region of the patient.

The subsequent sections of this chapter are organized as follows. Section 8.2 provides the technical overview of the proposed work providing details of the training and testing phases. The methodologies used are described in Sect. 8.3. The elaboration of the experimental results is given in Sect. 8.4 while Sect. 8.5 concludes the proposed work.

## 8.2 Technical Overview

The following figure in Fig. 8.1 depicts the schematic diagram of our proposed work.

The propose work utilizes three devices, namely

- 21-channel EEG signal acquisition device (manufactured by Nihon Kohden) [2–4, 8–12] to record the physiological signals from the brain.

**(a)**

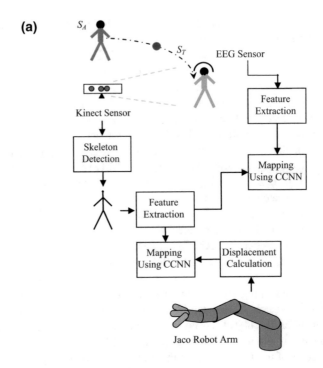

CCNN: Cascade-Correlation Neural Network
SA: Subject throwing the ball
ST: Target Subject

**(b)**

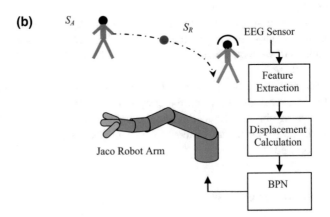

BPN: Bypass Network
SA: Subject throwing the ball
SR: Rehabilitative Patient

**Fig. 8.1  a**. Proposed system for rehabilitative applications: training phase. **b**. Proposed system for rehabilitative applications: testing phase

- Kinect 360 sensor [5–9, 11–14] (manufactured by Microsoft) to capture the body gesture of the subject in skeletal view by representing the body with twenty Cartesian joint co-ordinates in three dimensional space (3D).
- Jaco robot arm [1–4, 9] (manufactured by Kinova) is manipulated based on these two information.

The system is consists of two major parts: training phase and testing phase.

## 8.2.1 Training Phase (TP)

This phase consists of the steps need to undergo to train the artificial limb. This phase is bifurcated into two parts:

### 8.2.1.1 Training Phase Part 1 (TP1)

We have implemented the proposed work in such a fashion that the subject under consideration $(S_T)$ is asked to catch the ball thrown by a separate subject $(S_A)$. Now the target subject is wearing an EEG sensor [2–4, 8–12] and his/her body gestures are captured by the Microsoft's Kinect sensor [5–9, 11–14]. When the target subject notices a ball coming towards him/her then the occipital region in the brain gets activated. Now to catch the ball the parietal region of the brain planes how the body of the subject needs to be bent to catch that ball. Based on that signal from the parietal part of the brain, the motor cortex region instructs the hands of the subject to act accordingly. How the subject positioned his/her body to catch the ball, that gesture is accumulated by the Kinect sensor.

To achieve this three mappings are done between

- occipital signals $(OS_{TP}^T)$ and parietal signals $(PS_{TP}^T)$ to adapt weight vector $W_1$ using $CCLA_1$
- parietal signals $(PS_{TP}^T)$ and motor cortex signal $(MS_{TP}^T)$ to adapt weight vector $W_2$ using $CCLA_2$
- motor cortex signal $(MS_{TP}^T)$ and Kinetic data $(KD_{TP}^T)$ to adapt weight vector $W_3$ using $CCLA_3$.

where CCLA is cascade-correlation learning architecture [15, 16], *TP* is training phase and *T* denotes the target subject.

### 8.2.1.2 Training Phase Part 2 (TP2)

Now in the second sub part deals how the end effector of the Jaco robot arm needs to be placed in the exact position of the palm of the target subject such that the Jaco robot [1–4, 9] is able to catch the ball instead of the target subject. For this purpose

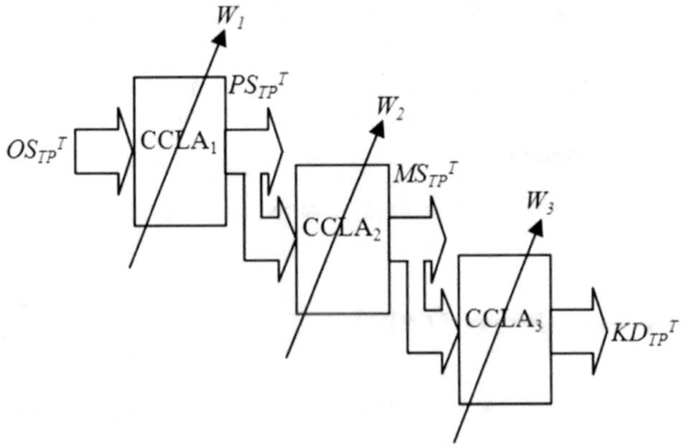

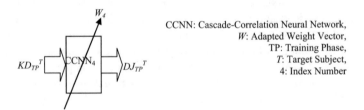

CCLA: Cascade-Correlation Learning Architecture,
W: Adapted Weight Vectors, TP: Training Phase,
T: Target Subject, 1-3: Index Numbers

**Fig. 8.2** Block diagram of training phase part 1

CCNN: Cascade-Correlation Neural Network,
W: Adapted Weight Vector,
TP: Training Phase,
T: Target Subject,
4: Index Number

**Fig. 8.3** Block diagram of training phase part 2

Kinetic data for the required gesture ($KD_{TP}^T$) and mapped with displacement of Jaco robot arm ($DJ_{TP}^T$) using cascade-correlation learning architecture(CCLA$_4$) [15, 16] and after successful run of CCLA the weight vector becomes $W_4$ (Figs. 8.2 and 8.3).

## 8.2.2 Testing Phase (EP)

After training the system with the above process, a bypass network is formed (BPN) such that when a ball is thrown to a rehabilitative patient ($S_R$) suffering from damage in the parietal and/or motor cortex region(s) then based on the signals obtained from occipital region ($OS_{EP}^R$), Jaco robot ($DJ_{EP}^R$) is instructed. Thus adapted weight vectors ($W_1$–$W_4$) are taken into account for construction of BPN. This network is termed as bypass network as here displacement of Jaco robot is possible by bypassing the signals from parietal, motor cortex regions and the Kinetic data (Fig. 8.4).

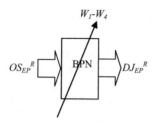

BPN: Bypass Network,
*W*: Adapted Weight Vector,
EP: Testing Phase,
R: Rehabilitative Patient,
1-4: Index Number

**Fig. 8.4**  Block diagram of testing phase

## 8.3  Methodologies Used

This section illustrates the procedure for feature extraction (FE) and mapping using cascade correlation learning architecture (CCLA).

### 8.3.1  Feature Extraction for EEG Signals

Features of EEG signal are represented as the basic primitives of the signal itself. In the present chapter, the significant information are extracted using Power spectral density (PSD) from the filtered EEG signal, which contains signal power distribution in the occipital lobe, parietal lobe and motor cortex. Mathematically, PSD may be defined as a Fourier Transform of the autocorrelation sequence of the time series, which is given below.

$$EEG(f) = \int_{t_1}^{t_2} eeg(t)e^{-2\pi ft}dt \tag{8.1}$$

$$PSD(f) = 2\frac{||EEG(f)||^2}{(t_2 - t_1)} \tag{8.2}$$

After FE, it is important to select the most significant EEG features without hampering the classification accuracy, if the feature dimension is high. PSD, being a high dimensional feature extractor, does not contain all important features. Therefore, we apply standard Principal component analysis (PCA) to select a fewer significant features from a large pool of features to train the neural network.

### *8.3.2　Feature Extraction for Kinetic Data (KD)*

As already stated, the Kinect sensor can represent a human body using twenty body joint co-ordinates (*JC*), but for this proposed work, the changes occur in the arms along with upper body (except head) co-ordinates have the major impact while catching a ball. Also the transition in those joints from initial state (when the subject first sees the ball coming) to goal state (when the subject is able to catch the ball) is irreverent as the degrees of freedom (DOF) for human arm ($dof_H$ = 27) [17] and Jaco arm ($dof_J$ = 7) including gripper [1] are not comparable. Thus we are only concerned about the difference between the joint co-ordinates between initial state ($IS^{KD}$) and goal state ($GS^{KD}$). The required joints in 3D (*x*, *y* and *z*) are spine (*S*), shoulder center (*SC*), shoulder left (*SL*), elbow left (*EL*), wrist left (*WL*), hand left (*HL*), shoulder right (*SR*), elbow right (*ER*), wrist right (*WR*) and hand right (*HR*). The extracted feature vector ($FV_{KD}$) is given in (8.3).

$$
FV_{KD} = \begin{bmatrix}
S_x^{GS} - S_x^{IS} & S_y^{GS} - S_y^{IS} & S_z^{GS} - S_z^{IS} \\
SC_x^{GS} - SC_x^{IS} & SC_y^{GS} - SC_y^{IS} & SC_z^{GS} - SC_z^{IS} \\
SL_x^{GS} - SL_x^{IS} & SL_y^{GS} - SL_y^{IS} & SL_z^{GS} - SL_z^{IS} \\
EL_x^{GS} - EL_x^{IS} & EL_y^{GS} - EL_y^{IS} & EL_z^{GS} - EL_z^{IS} \\
WL_x^{GS} - WL_x^{IS} & WL_y^{GS} - WL_y^{IS} & WL_z^{GS} - WL_z^{IS} \\
HL_x^{GS} - HL_x^{IS} & HL_y^{GS} - HL_y^{IS} & HL_z^{GS} - HL_z^{IS} \\
SR_x^{GS} - SR_x^{IS} & SR_y^{GS} - SR_y^{IS} & SR_z^{GS} - SR_z^{IS} \\
ER_x^{GS} - ER_x^{IS} & ER_y^{GS} - ER_y^{IS} & ER_z^{GS} - ER_z^{IS} \\
WR_x^{GS} - WR_x^{IS} & WR_y^{GS} - WR_y^{IS} & WR_z^{GS} - WR_z^{IS} \\
HR_x^{GS} - HR_x^{IS} & HR_y^{GS} - HR_y^{IS} & HR_z^{GS} - HR_z^{IS}
\end{bmatrix}
\tag{8.3}
$$

### *8.3.3　Feature Extraction for Jaco Displacements*

Jaco robot arm is manufactured by Kinova and is a 6-axis robotic manipulator with a three fingered gripper to grip an object (here ball). The arm has 7 degree of freedom (DOF) [1] with a maximum reach of 90 cm and has maximum linear speed of 20 cm/s. For displacement of Jaco robot (*DJ*) from initial state ($IS^{DJ}$) to goal state ($GS^{DJ}$), we have to specify the following feature vector $FV_{DJ}$ using (8.4).

$$
FV_{DJ} = \begin{bmatrix} J_1^\theta & J_2^\theta & J_3^\theta & J_4^\theta & J_5^\theta & J_6^\theta \end{bmatrix}
\tag{8.4}
$$

where 1–6 indicates the joint (*J*) number of the Jaco robot and $\theta$ denotes the desired angle of that respective joint needs to be rotated to act accordingly. The value of $\theta$ can be positive or negative based on whether it is clockwise or anticlockwise.

## 8.3.4 Mapping Using Cascade-Correlation Learning Architecture

In the view of real-world problem, computational cost is the main bottleneck of the existing artificial neural network based learning algorithm (e.g., back-propagation). The former limitation is overcome by employing the Cascade-Correlation Learning Architecture (CCLA) [15] (as shown in Fig. 8.5), which is an artificial neural network based algorithm, begins with minimum input and output nodes (iteration 1 in Fig. 8.6). Gradually the network trains and adds hidden layers to create a multi-layered neural network. All the hidden neurons have fixed weight (square connection) in the input side as shown in Fig. 8.5 and only the output side weights (crossed connections) are trained. The bias is fixed at +1. Unlike back-propagation [8], the CCLA trains the weights directly connected to the outputs, which results in incredible speed-up as compared to the back-propagation. Also CCLA creates promising high-order feature detectors.

The CCLA is a twofold idea:

- *One is the addition of cascade architecture to add fixed hidden layer one at a time to the network.*
- *Another is the learning algorithm to install a hidden layer with a symmetric Sigmoidal activation function such that the correlation between the new network's output and the residual error signal (to be minimized) is maximized.*

Training the weights directly connected to the output neurons, in CCLA, motivates to employ the single-layered learning algorithms, e.g., the Widrow-Hoff delta learning rule, the Perceptron learning and the resilient propagation (RPROP) algorithm [16]. In this chapter, RPROP is employed for training.

New hidden layer is created by selecting a candidate unit (e.g.., sigmoidal function, Gaussian function, radial basis function, etc.), which can be connected to all the pre-existing input, output and hidden layers. A number of passes is done to adjust and fix the input weights of the candidate unit. Finally the output of the candidate unit is connected to the network.

The input weights of the candidate unit are adjusted with an aim to maximize the sum over all the output units' covariance given in (8.5),

$$C_v = \sum_o | \sum_p (V_p - \bar{V})(E_{p,o} - \bar{E})| \tag{8.5}$$

where, $V$ be the candidate unit's value, $E_{o,p}$ be the residual output error observed at output unit $o$ in the $p$th training pattern and $\bar{V}, \bar{E}_o$ are the average values of $V, E_o$ respectively over all the training pattern. In CCLA, like back-propagation $\frac{\partial C_v}{\partial w_i}$ is computed in (8.6) to maximize $C_v$ where, $w_i$ is the $i$th candidate units incoming weight.

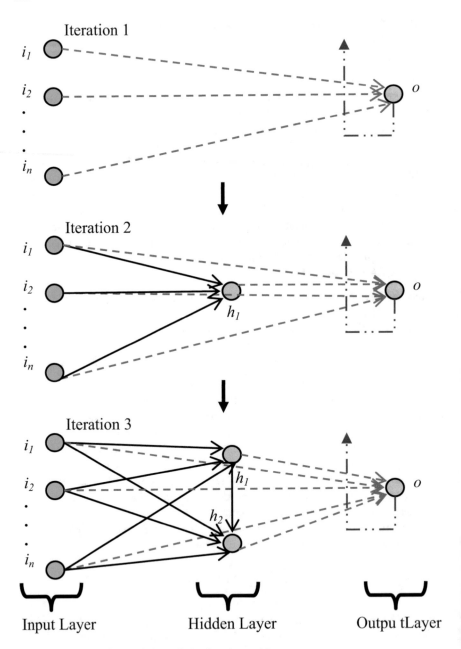

**Fig. 8.5** Structure of cascaded correlation learning architecture

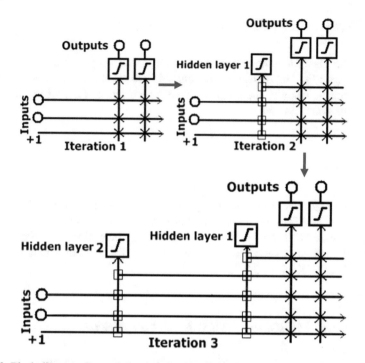

**Fig. 8.6** Block diagram of cascaded correlation learning architecture

$$\frac{\partial C_v}{\partial w_i} = \sum_{p,o} \sigma_o (E_{p,o} - \bar{E}_o) f'_p I_{i,p} \tag{8.6}$$

where, $\sigma_o$ be the correlation sign between the candidate value and output $f'_p$ be the first derivative of the candidate unit's activation function with respect to the sum of its inputs for pattern $p$ and $I_{i,p}$ refers the input revived by the candidate unit from $i$th candidate for pattern $i$. A gradient ascent maximization $C_v$ follows after computing $\frac{\partial C_v}{\partial w_i}$ for each incoming connection. The new candidate unit is connected as an fixed and active part of the network as $C_v$ stops improving. It is possible to employ a set of candidate units with randomly initialized input weights, instead of employing single candidate unit to install hidden layers in the network. When a set of candidate units are employed, they receive the same input, residual error and output, because the candidate units' outputs are not connected to the active network. Employing a set of candidate units is advantageous in two ways: one it reduces the chance of installing an useless candidate, which got stuck during training and another is it can speed-up the training because of the simultaneous wide exploration of the weight-space.

## 8.4   Experimental Results

This section elaborates the experimental setup along with subject details, signals analysis from EEG and Kinect, movements of Jaco robots. The section also shows the performance study of the proposed system with other existing literatures using metrics and statistical test.

### 8.4.1   Setup

The experiment has been performed at Artificial Intelligence Lab, Jadavpur University, where the framework includes a Kinect sensor (Fig. 8.7a) and a 21-channel wired EEG device (Fig. 8.7b). It has also been clear from Fig. 8.7b that Kinect data and EEG signals are recorded on two different computers with 8 GB RAM with CPU clock of 3.4 GHz.

### 8.4.2   Data Acquisition Using EEG and Kinect

We have created 3 datasets with 24 healthy subjects (15 men and 9 women). In each dataset, equal number of subjects are taken where gender ratio is 5:3 and for these 3 datasets the age groups are $(25 \pm 2\text{years})$, $(28 \pm 2\text{years})$ and $(30 \pm 2\text{years})$ in the training experiments. Among 24, 12 subjects, 6 men and 6 women have been participated for the testing sessions.

#### 8.4.2.1   Using EEG

In this experiment, EEG signals from electrode positions: P3, P4, Pz, C3, C4, Cz, O1 and O2 are recorded by using a 21-channel stand-alone EEG device having a sampling rate of 200 Hz and resolution of 100 μV (Fig. 8.7b). Each subject is instructed to throw the ball to opponent from eleven different positions. Figure 8.8a, b shows a clear view of EEG acquisition system used here and the signal pattern from the above electrode positions during the experiment.

Initially the samples taken for eleven instances are 1400, 3800, 1600, 1800, 1600, 1600, 1600, 1800, 1800, 2200, 2000 respectively with sampling rate 200 Hz. EEG signals captured during the above mentioned eleven instances are pre-processed and further analyzed. Figures 8.9, 8.10 and 8.11 present the raw EEG signals acquired from occipital, parietal and motor cortex regions respectively during two different instances.

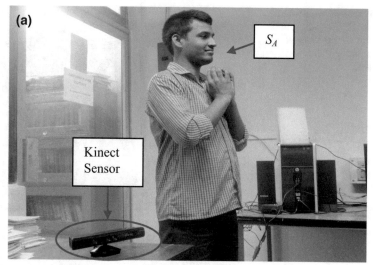

Towards $S_A$ end

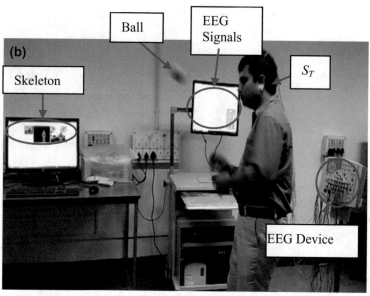

Towards $S_R$ end

**Fig. 8.7** Total setup of the proposed work for rehabilitative applications

Three following observations are noted from Figs. 8.9, 8.10 and 8.11.

- *There is a close parity in signal patterns recorded from the same set of electrodes for two different instances.*

**Fig. 8.8** **a**. EEG acquisition system, **b**. EEG signal from eight electrode positions is recorded during the experiment

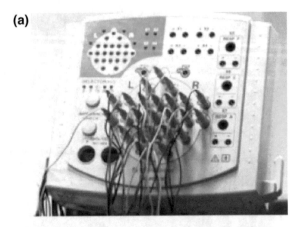

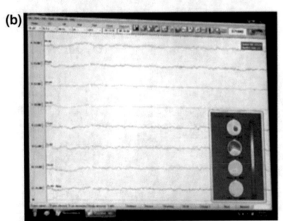

- *For each brain lobe (i.e., occipital, parietal and motor cortex), changes in signal pattern from their usual nature near about the same samples ensure the correlation between three brain lobes for a particular given instance.*
- *Changes in EEG pattern in different lobes validate the well-known EEG signal modalities. From Figs. 8.9 and 8.10, sudden positive rise in signal amplitude ensures the liberation of P300 modality from occipital and parietal lobes, whereas from Fig. 8.11 de-synchronization of EEG signal from the usual pattern at the start of motor action and again synchronized with the original pattern after completion of the motor activity ensures the association of event-related desynchronization/synchronization (ERD/ERS) modality during movement-related tasks.*

Since, the raw EEG signals contain noise and eye-blinking artifacts, following two steps have been performed to make the EEG signal artifact-free. First, we designed a band pass filter having suitable pass band frequencies to filter the line noise. Here, we use a Chebyshev type 2 filter having variable pass band frequency.

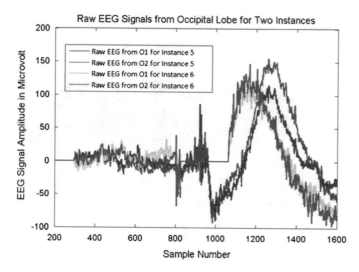

**Fig. 8.9** Raw EEG signals recorded from O1 and O2 electrodes during 5th and 6th instances

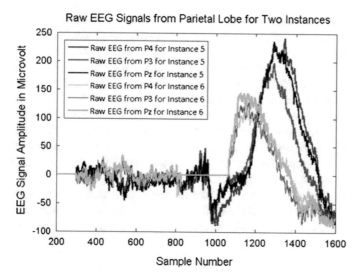

**Fig. 8.10** Raw EEG signals recorded from P3, P4 and Pz electrodes during 5th and 6th instances

For visual signal, pass band of the above filter is selected from 3 to 13 Hz in order to pass necessary information involved in theta (3–7 Hz) and alpha (7–13 Hz) band. For motor imagery and execution, pass band frequency is so chosen that it includes sensori-motor (7–13 Hz) and beta (13–30 Hz) rhythm. Second, we perform Independent component analysis (ICA) [18] to remove eye-blinking artifacts

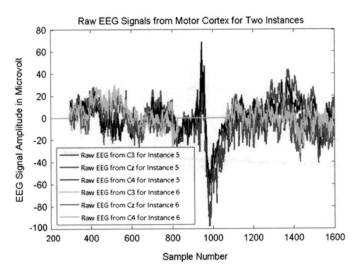

**Fig. 8.11** Raw EEG signals recorded from C3, C4 and Cz electrodes during 5th and 6th instances

from the captured EEG signals. In this technique, the scalp-components having artifacts are rejected and the remaining artifact-free components are selected.

Once pre-processing of EEG signal is done, features are extracted from the filtered EEG samples taken at different experimental instances. Figures 8.12, 8.13 and 8.14 present PSD feature discrimination of EEG signals recorded from occipital, parietal and motor cortex regions respectively during eleven different instances. It is important to note from Figs. 8.12, 8.13 and 8.14 that although PSD extracts more than several hundreds of features from each lobe, all features do not have significant values. For example, beyond 80th feature (for occipital EEG) and beyond 260th feature (for parietal and motor cortex EEG), we get redundant features. Another interesting fact is that for three brain regions, almost all instances have overlapped features except for instance 2 and hence a few features have capability to discriminate all eleven instances. To overcome this problem, we apply standard PCA algorithm [8] which selects 12 features that correctly discriminate all eleven instances.

### 8.4.2.2  Using Kinect

The Kinect sensor basically has three sensors responsible for three types of outputs.

- *RGB (red, green and blue) video is obtained from RGB camera.*
- *Depth map is obtained from IR (infrared) emitter and IR receiver.*
- *Twenty body joint co-ordinates in 3D space is obtained based on RGB and Depth data using software development kit (SDK).*

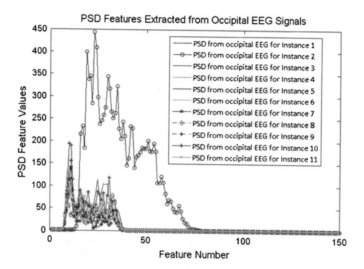

**Fig. 8.12** PSD features extracted from occipital lobe during eleven instances

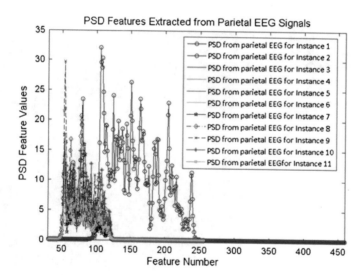

**Fig. 8.13** PSD features extracted from parietal lobe during eleven instances

The details of the outputs obtained from Kinect sensor is depicted in Fig. 8.15. We have used SDK version 1.6 and recorded the initial state and goal state skeleton data which is given in Table 8.1.

The RGB images taken by a separate camera placed perpendicular and parallel to Kinect to provide better visualization for the 1st instance for 5th subject is shown in

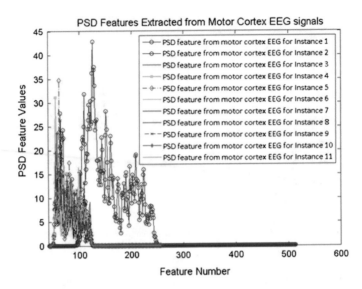

**Fig. 8.14** PSD features extracted from motor cortex during eleven instances

**Fig. 8.15** Kinect sensor
displaying three sensors,
along with the outputs
obtained from it

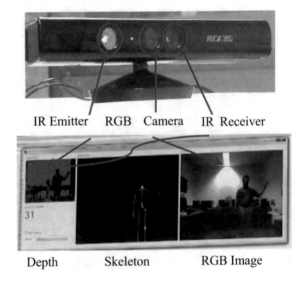

Fig. 8.16. The figure elaborates how the gesture for subject $S_T$ changes through $IS^{KD} \rightarrow INS_1^{KD} \rightarrow INS_2^{KD} \rightarrow INS_3^{KD} \rightarrow GS^{KD}$. The corresponding $FV_{KD}$ obtained is given in (8.2).

**Table 8.1** The eleven instances taken for 5th Subject

| Initial State ($IS^{KD}$) |
| --- |

| Eleven Goal States ($GS^{KD}$) |
| --- |

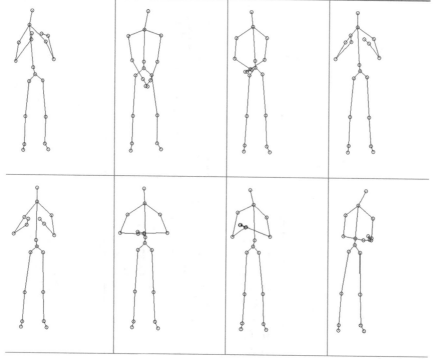

(continued)

**Table 8.1** (continued)

Initial State ($IS^{KD}$)

Eleven Goal States ($GS^{KD}$)

  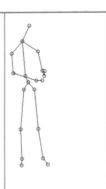

$$FV_{KD} = \begin{bmatrix} 0.0094 & -0.0755 & -0.0498 \\ 0.0175 & -0.0515 & -0.0179 \\ 0.0177 & -0.0598 & -0.0548 \\ 0.0530 & -0.0332 & -0.0296 \\ 0.1177 & 0.4650 & -0.0850 \\ 0.1079 & 0.5912 & -0.1334 \\ -0.0023 & -0.0668 & -0.0471 \\ 0.0061 & -0.0070 & -0.1010 \\ -0.09404 & 0.4986 & -0.1082 \\ -0.0945 & 0.6190 & -0.1615 \end{bmatrix} \qquad (8.7)$$

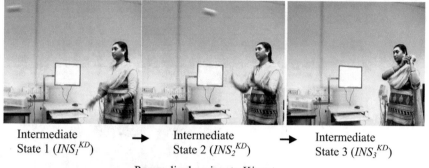

Intermediate
State 1 ($INS_1^{KD}$) ➡ Intermediate
State 2 ($INS_2^{KD}$) ➡ Intermediate
State 3 ($INS_3^{KD}$)

Perpendicular view to Kinect

Initial State ($IS^{KD}$) ➡ Goal State ($G^{KD}$)

Parallel view to Kinect

**Fig. 8.16** RGB images for 1st instance for 5th subject $S_T$

## 8.4.3 Parameters Settings for CCLA

Table 8.2 shows the design of CCLA. The parameters set for RPROP [16] are $\Delta_{min} = 0.0001$, $\Delta_{ini} = 0.01$, $\Delta_{max} = 0.5$, $\eta^- = 0.5$ and $\eta^+ = 1.2$.

## 8.4.4 Displacement of Jaco Robot

The Jaco robot arm as shown in Fig. 8.17 is suitable for a person with disability of the upper arm (parietal and/or motor cortex) and can be attached to a wheelchair or somewhere else to assist the disabled person [9].

**Table 8.2** Table XX Design of all the CCLA blocks

| Parameters | CCLA$_1$ | CCLA$_2$ | CCLA$_3$ | CCLA$_4$ |
|---|---|---|---|---|
| Input node | $2 \times 12$ | $3 \times 12$ | $3 \times 12$ | $10 \times 3$ |
| Output node | $3 \times 12$ | $3 \times 12$ | $10 \times 3$ | $1 \times 6$ |
| Passes | 500 | 800 | 750 | 400 |
| Candidate units | 40 | 60 | 50 | 30 |

**Fig. 8.17** Jaco robot arm

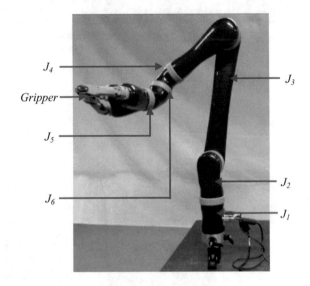

The RGB images taken by a separate camera placed parallel to Kinect (as Kinect's RGB camera has only 8-bit VGA resolution) for better view and picture quality for the 1st instance for 10th subject is shown in Fig. 8.18. The figure elaborates how the Jaco moves for subject $S_R$ changes through $IS^{DJ} \rightarrow INS_1^{DJ} \rightarrow INS_2^{DJ} \rightarrow GS^{DJ}$. The exact movements occur in different joints for Jaco whole accommodating the desired movement is given in Table 8.2.

### 8.4.5  Performance Analysis

We have compared CCLA performances with back propagation neural network (BPNN) [8], feed forward neural network (FFNN) [19], adaptive neural fuzzy inference (ANFI) [3], ensemble classifier using binary tree (ECBT) [20], linear support vector machine (LSVM) [20], support vector machine with radial basis function (RBFSVM) [21], AdaBoost-Support Vector (ASVM) [4], k-nearest neighbor (kNN) using the metrics including accuracy, precision, sensitivity,

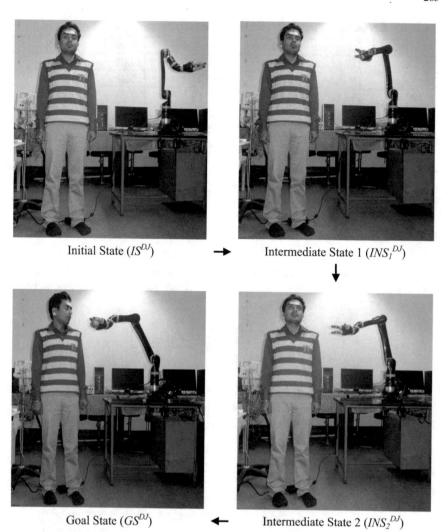

Initial State ($IS^{DJ}$)   →   Intermediate State 1 ($INS_1^{DJ}$)

↓

Goal State ($GS^{DJ}$)   ←   Intermediate State 2 ($INS_2^{DJ}$)

**Fig. 8.18** RGB images for 1st instance for 10th rehabilitative patient $S_R$

specificity and F1 score. The average results for 3 datasets for these metrics are given in Table 8.3.

All the experiments are performed on a machine with Intel(R) core(TM) i7-4790 processor and 8 GB of DDR2-memory. For the training stage, the codes are written in Matlab R2012b and Microsoft Visual Studio 2010 version is employed to test the experiments in real-time with Jaco arm.

**Table 8.3** Comparison of proposed work with existing literatures

| Algorithms | Accuracy | Precision | Sensitivity | Specificity | F1 Score |
|---|---|---|---|---|---|
| CCLA | 87.65 | 86.74 | 84.52 | 87.02 | 88.83 |
| BPNN | 80.92 | 79.47 | 81.91 | 82.22 | 80.30 |
| FFNN | 77.72 | 75.07 | 76.18 | 77.00 | 74.93 |
| ANFI | 83.46 | 80.16 | 83.08 | 82.66 | 83.17 |
| ECBT | 71.49 | 73.08 | 75.12 | 72.92 | 71.74 |
| LSVM | 58.84 | 59.86 | 59.29 | 61.49 | 60.71 |
| RBFSVM | 71.05 | 72.75 | 71.27 | 72.71 | 73.98 |
| ASVM | 73.03 | 72.89 | 72.93 | 71.22 | 73.71 |
| kNN | 68.57 | 68.75 | 69.52 | 71.31 | 67.68 |

### 8.4.6  Friedman's Statistical Test

The best of all the $c$ algorithms, i.e., $1 \leq c \leq C$, is given rank 1 and the worst is given rank $D$. The average ranking obtained by the $c$-th algorithm over all $d$ $(1 \leq d \leq D)$ datasets is $R_c$.

$$\chi^2 = \frac{12D}{C(C+1)} \left[ \sum_{c=1}^{C} R_c^2 - \frac{C(C+1)^2}{4} \right] \tag{8.8}$$

Here, $D = 3$ and $C = 9$. In Table 8.4, it is shown that the null hypothesis has been rejected, as $\chi_F^2 = 23.64$ is greater than 15.51, the critical value of $\chi^2$ distribution for $C - 1 = 8$ DOF with 95% accuracy.

**Table 8.4** Performance analysis using Friedman's test

| Algorithm | Dataset 1 | Dataset 2 | Dataset 3 | $R_c$ | $\chi^2$ |
|---|---|---|---|---|---|
| CCLA | 1 | 1 | 1 | 1.00 | 23.64 |
| BPNN | 3 | 2 | 3 | 2.67 | |
| FFNN | 5 | 4 | 4 | 4.33 | |
| ANFI | 2 | 3 | 2 | 2.33 | |
| ECBT | 6 | 6 | 6 | 6.00 | |
| LSVM | 9 | 9 | 9 | 9.00 | |
| RBFSVM | 7 | 7 | 7 | 7.00 | |
| ASVM | 4 | 5 | 5 | 4.67 | |
| kNN | 8 | 8 | 8 | 8.00 | |

## 8.5 Conclusion

The proposed work implements a novel technique for movement of artificial limb based on simultaneous signals from EEG and gesture. In the first training phase, the four mappings are done between three (occipital, parietal and motor cortex) brain signals and skeleton obtained from Kinect sensor. In the next training phase, another mapping is done to direct the Jaco robot based on skeleton. All the four mapping are done using dedicated cascade correlation learning architecture. In the testing phase, a bypass network is created to employ the Jaco robot arm for assisting the rehabilitative patient with damage in parietal and/or motor cortex region(s). This system has several applications in the development of rehabilitative aids.

The proposed system uses cascade correlation learning architecture as this network shows much better performance dealing with the weight adaptation and randomness when compare with other neural networks including back propagation neural network. For the real time implementation of the proposed work, a proper selection of network has vital importance. The system shows a high accuracy of 87.65% for rehabilitative patients.

## References

1. V. Maheu, J. Frappier, P.S. Archambault, F. Routhier, Evaluation of the JACO robotic arm: Clinico-economic study for powered wheelchair users with upper-extremity disabilities, in *Rehabilitation Robotics (ICORR), 2011 IEEE International Conference on*, 2011, pp. 1–5
2. L. Bougrain, O. Rochel, O. Boussaton, L. Havet, From the decoding of cortical activities to the control of a JACO robotic arm: a whole processing chain, *arXiv Prepr. arXiv1212.0083* (2012)
3. S. Bhattacharyya, D. Basu, A. Konar, D.N. Tibarewala, Interval type-2 fuzzy logic based multiclass ANFIS algorithm for real-time EEG based movement control of a robot arm. Rob. Auton. Syst. **68**, 104–115 (2015)
4. S. Bhattacharyya, A. Konar, D.N. Tibarewala, Motor imagery, P300 and error-related EEG-based robot arm movement control for rehabilitation purpose. Med. Biol. Eng. Comput. **52**(12), 1007–1017 (2014)
5. B. Wang, C. Yang, Q. Xie, Human-machine interfaces based on EMG and Kinect applied to teleoperation of a mobile humanoid robot, in *Intelligent Control and Automation (WCICA), 2012 10th World Congress on* (2012), pp. 3903–3908
6. G. Du, P. Zhang, Markerless human–robot interface for dual robot manipulators using Kinect sensor. Robot. Comput. Integr. Manuf. **30**(2), 150–159 (2014)
7. A. Csapo, E. Gilmartin, J. Grizou, J. Han, R. Meena, D. Anastasiou, K. Jokinen, G. Wilcock, Multimodal conversational interaction with a humanoid robot, in *Cognitive Infocommunications (CogInfoCom), 2012 IEEE 3rd International Conference on* (2012), pp. 667–672
8. S. Datta, A. Khasnobish, A. Konar, D.N. Tibarewala, A.K. Nagar, EEG based artificial learning of motor coordination for visually inspired task using neural networks, in *Neural Networks (IJCNN), 2014 International Joint Conference on* (2014) pp. 3371–3378
9. A. Frisoli, C. Loconsole, D. Leonardis, F. Banno, M. Barsotti, C. Chisari, M. Bergamasco, A new gaze-BCI-driven control of an upper limb exoskeleton for rehabilitation in real-world tasks, Syst. Man, Cybern. Part C Appl. Rev. IEEE Trans. **42**(6), 1169–1179 (2012)

10. R. Roy, A. Konar, D. N. Tibarewala, EEG driven artificial limb control using state feedback PI controller," in *Electrical, Electronics and Computer Science (SCEECS), 2012 IEEE Students' Conference on* (2012), pp. 1–5
11. N. A. da Silva, R. Maximiano, H. A. Ferreira, Hybrid Brain Computer Interface Based on Gaming Technology: An Approach with Emotiv EEG and Microsoft Kinect, in *XIII Mediterranean Conference on Medical and Biological Engineering and Computing 2013* (2014), pp. 1655–1658
12. R. Scherer, J. Wagner, G. Moitzi, G. Muller-Putz, Kinect-based detection of self-paced hand movements: enhancing functional brain mapping paradigms, in *Engineering in Medicine and Biology Society (EMBC), 2012 Annual International Conference of the IEEE* (2012), pp. 4748–4751
13. I. Siradjuddin, L. Behera, T. M. McGinnity, S. Coleman, A position based visual tracking system for a 7 DOF robot manipulator using a Kinect camera, in *Neural Networks (IJCNN), The 2012 International Joint Conference on* (2012), pp. 1–7
14. G. Csaba, L. Somlyai, Z. Vámossy, Differences between Kinect and structured lighting sensor in robot navigation, in *Applied Machine Intelligence and Informatics (SAMI), 2012 IEEE 10th International Symposium on* (2012), pp. 85–90
15. S. E. Fahlman, C. Lebiere, *The cascade-correlation learning architecture.* (1989)
16. M. Riedmiller, H. Braun, A direct adaptive method for faster backpropagation learning: The RPROP algorithm," in *Neural Networks, 1993., IEEE International Conference on* (1993), pp. 586–591
17. G. ElKoura, K. Singh, Handrix: animating the human hand, in *Proceedings of the 2003 ACM SIGGRAPH/Eurographics symposium on Computer animation* (2003), pp. 110–119
18. T.-W. Lee, *Independent component analysis.* (Springer, 1998)
19. M. Leena, K. Srinivasa Rao, B. Yegnanarayana, Neural network classifiers for language identification using phonotactic and prosodic features, in *Intelligent Sensing and Information Processing, 2005. Proceedings of 2005 International Conference on* (2005), pp. 404–408
20. S. Saha, S. Datta, A. Konar, R. Janarthanan, A study on emotion recognition from body gestures using Kinect sensor, in *Communications and Signal Processing (ICCSP), 2014 International Conference on* (2014), pp. 56–60
21. S. Saha, S. Ghosh, A. Konar, A. K. Nagar, Gesture Recognition from Indian Classical Dance Using Kinect Sensor, in *Computational Intelligence, Communication Systems and Networks (CICSyN), 2013 Fifth International Conference on* (2013), pp. 3–8

# Chapter 9
# Conclusions and Future Directions

This chapter provides a thorough review of the principles and concepts narrated in the book, highlighting significant results and conclusions arrived at earlier. In other words, the chapter combines important results to offer more generic conclusions for future researchers in gesture recognition. Possible direction of future research is also indicated in this chapter.

## 9.1  Self-Review of the Book

The book offers both a theoretical overview and experiments on gesture recognition. Gestural data have been collected from images and/or Microsoft Kinect data files for real-world problems of diverse interests. Among the interesting problems addressed in the book, the main focus lies in recognition of primitive dance postures, healthcare and gaming. All the three applications have been illustrated in greater detail.

Chapter 1 covers foundations in gesture recognition. It aims at projecting the principle of gesture recognition through plenty of illustrations so as to help a beginner grasp the subject without difficulty. The chapter includes sensory modalities in gesture recognition from two angles: (i) direct real data from Microsoft Kinect machine and (ii) raw images containing gestural information. The first modality is used for healthcare, e-learning of dances and gaming. The second modality has been used to develop applications for e-learning of Ballet only. As a whole, the chapter outlined the principles of feature extraction from gestural data and demonstrates the scope of pattern classification to classify gestural information into primitive classes, having meaningful attributes. The chapter also provides a survey of the existing works with special emphasizing on the scope of Finite State Machines, Hidden Markov model and Particle Filter in gesture recognition.

Chapter 2 dealt with a very interesting and useful problem on e-learning of ballet from videos (containing the dance) with a provision for automatic pin-pointing of

© Springer International Publishing AG 2018
A. Konar and S. Saha, *Gesture Recognition*, Studies in Computational
Intelligence 724, DOI 10.1007/978-3-319-62212-5_9

the errors performed by the learner during her practise session. A fast but accurate method to detect primitive ballet gestures from captured images has been presented in this chapter. The attractive part of the classification algorithm lies in extracting minimal features, here Radon transform and Euler number only, to classify ballet primitives into 17 classes.

Although Chap. 2 is good enough to classify accurate ballet gestures, for novice ballet learners the accuracy in the gestural pattern usually is found to have a deviation from their ideal ones. This inspired us to develop Chap. 3, where we capture the uncertainty in ballet dance by the logic of fuzzy sets and proposed a solution to primitive gesture classification in presence of uncertainty by fuzzy rule based approach.

The newly developed Microsoft's Kinect sensor has opened up an innovative horizon to recognize human body using three-dimensional body joints. In Chap. 4, a novel gesture driven car racing game control is proposed using Mamdani type-I fuzzy inference. The technique is very useful for providing better gaming experience to the gamers, as they can get the sensation of turning virtual steering wheel by moving their arms synergistically. The speed of the virtual racing car can be controlled by the same way in actual cars using the relative displacements of the two feet.

Now-a-days most of families are nuclear and the elderly people are forced to live alone in their homes without any supervision. For these types of homes, a system has been developed for home monitoring of elderly people in the fifth chapter. Certain gestures shown by the elderly people indicate suffering from joint and muscle pains are considered here. The features acquired from these gestures tend to have wider uncertainties when data is gathered from a large number of subjects. To tackle this, type-2 fuzzy set is introduced in this chapter.

For the purpose of e-learning, Kinect sensor has tremendous scope. Thus in Chap. 6, Kinect sensor based dance e-leaning technique is proposed. The basic idea of geometry is used to create five planes based on five sections of human body. The relative movements of these planes are considered as features. The gestures are classified using probabilistic neural network.

In the Chap. 7, a novel idea of dance composition is proposed using differential evolution algorithm. The system presented here takes the knowledge of large amount of dance gestures already stored in the database. The gestures are acquired using Microsoft Kinect sensor.

The Chap. 8 focuses on providing aid to the rehabilitative patients. For the assistance, an artificial limb is directed based on brain signals and body gestures. To achieve the timing constraint parameter of real-time system, cascade correlation learning architecture is used in this chapter.

## 9.2  The Future of Gesture Recognition

Gesture recognition is receiving increasing attention from diverse domains of applications. One important problem in defence sector is to recognize a (foreign) enemy in the troop of a country, particularly if he is in disguise. One way to recognize him is to study his behavioural interaction with soldiers and gesture-analysis [1, 2]. This requires taking multiple snaps of the gestural patterns of the person and their subsequent analysis.

The second interesting application of gesture recognition is to analysis the movement pattern of people for their gait analysis and thus to recognize the person. In criminal investigation problem, gait analysis has an important role to identify suspects from the verbal description of their movement patterns from the eye-witness. In a recent US patent [3], the patent-holders indicated gait analysis of a subject by using two Microsoft Kinect machines to cover a long range of the subject's fast and slow walking patterns.

Game-teaching by copying gesture of a good-performer in outdoor games like cricket/tennis/basket ball is an interesting application of gesture recognition [4, 5]. This can be realized first by storing gestural patterns for the successful moves of the player and subsequently asking the children to copy his movement patterns in the subsequent games.

In the theoretical side, there exists immense scope for gesture analysis by 3D imaging. The idea lies in collecting a set of 3D points resembling a gesture. Prediction of gestures when part of the data points are missing due to occlusion is an important research arena [6, 7]. Traditional prediction algorithms, such as Kalman filter and more recent technique, such as Particle Filter, would have a great role to play in the present prediction problem.

Time-motion study of the workers, particularly industrial labours, is an important aspect to examine their performance [8]. In late 1990's, time-motion study was performed by manually examining the time spent in completing a job. One modern approach of time-motion study is to measure the speed of individual body junctions involved in executing a complex task. The Microsoft Kinect may be employed to acquire the junction coordinates at different phases of a mechanical machining task. The speed of different junctions and the rate of interactions between any two junctions may be defined important metrics to compare two people engaged in similar tasks.

The last but not the least application area that needs special mention is gesture-induced robotic surgery [9, 10]. This is important for senior surgeons who sometimes are tired after performing several surgeries on the same date. Alternatively, for finer level of surgery, assistive robotics is a challenging medium to handle the problem. In assistive robotics, the doctor may command the robot by his gesture to execute specific tasks. Surgical robotics has a good future in the coming decades.

# References

1. C.J. Cohen, K.A. Scott, M.J. Huber, S.C. Rowe, F. Morelli, Behavior Recognition Architecture for Surveillance Applications, in *Applied Imagery Pattern Recognition Workshop, 2008. AIPR'08. 37th IEEE* (2008), pp. 1–8
2. C.J. Cohen, F. Morelli, K.A. Scott, A, Surveillance System for the Recognition of Intent within Individuals and Crowds, in *2008 IEEE Conference on Technologies for Homeland Security,* (2008), pp. 559–565
3. K. Chakravarty, A. Sinha, D. Das, R. Banerjee, A. Konar, S. Dutta, Identification of People Using Multiple Skeleton Recording Devices. Google Patents, 08 Dec 2015
4. M.-H. Yang, N. Ahuja, M. Tabb, Extraction of 2d motion trajectories and its application to hand gesture recognition. Pattern Anal. Mach. Intell. IEEE Trans. **24**(8), 1061–1074 (2002)
5. V. Bloom, D. Makris, V. Argyriou, G3d: A Gaming Action Dataset And Real Time Action Recognition Evaluation Framework, in *2012 IEEE Computer Society Conference on Computer Vision and Pattern Recognition Workshops (CVPRW),* (2012), pp. 7–12
6. Y. Wu, T.S. Huang, Vision-based Gesture Recognition: A review, in *Urbana*, vol 51 (1999), p. 61,801
7. S.J. McKenna, S. Jabri, Z. Duric, A. Rosenfeld, H. Wechsler, Tracking groups of people. Comput. Vis. Image Underst. **80**(1), 42–56 (2000)
8. R.M. Barnes, R.M. Barnes, in *Motion and time study*, vol. 84, (Wiley New York, New York, 1958)
9. G.H. Ballantyne, Robotic surgery, telerobotic surgery, telepresence, and telementoring. Surg. Endosc. **16**(10), 1389–1402 (2002)
10. A.R. Lanfranco, A.E. Castellanos, J.P. Desai, W.C. Meyers, Robotic surgery: a current perspective. Ann. Surg. **239**(1), 14–21 (2004)

# Index

## A

Acceleration, 117, 118
Accelerometer, 22
Accuracy, 35, 135, 138, 147, 190
AdaBoost-support vector, 264
Adaptive bee colony algorithm, 239
Adaptive boosting, 22, 54, 176
Adaptive neural fuzzy inference, 264
American sign language, 20
Angle, 137
Angle of steering, 121
Artifact, 258
Artificial limb, 29, 243
Artificial neural network, 176
Autocorrelation, 249
Average error rate, 178

## B

Background subtraction, 19
Back propagation neural network, 14, 205,
    208, 264, 267
Baum-Welch algorithm, 25
Bayer color filter, 17
Bayesian model, 10, 24
Behavioural interaction, 271
Binary tree, 208, 264
Blending factor, 54
Bonferroni-Dunn test, 185
Bounded region, 66
Brake, 117, 125, 127
Branch point, 74
Bypass network, 245, 248

## C

Cascade correlation learning architecture, 14,
    29, 243, 249, 267, 270
Cerviacal Spondylosis, 162
Chain code, 66
Chebyshev type 2 filter, 256

Children tantrum, 136
Chi square distribution, 56
Choreograph, 226
Circumplex model, 199
Classification, 138
Competitive layer, 206
Computational complexity, 187
Connected component, 38, 74
Correlation coefficient, 37
Correlation learning architecture, 14, 29
Covariance, 251
Critical difference, 185
Critical value, 56, 181
Crossover, 16

## D

Dance, 16, 20, 27, 44, 195
Decay rate, 149, 152
Degree of freedom, 132, 144, 181
Degree of membership, 152
Degree of uncertainty, 137
Depth, 17
De-synchronization, 256
Differential evolution, 16, 29, 225
Dilation, 38
Disorder, 135, 168
Drastic product, 85
Driving, 117
Driving gesture, 2, 132
Dynamic, 1
Dynamic Bayesian network, 22
Dynamic time warping, 21

## E

Edge, 7
Edge orientation histogram, 20
E-learning, 26, 195, 198
Electroencephalography, 29, 243
EMG sensor, 245

© Springer International Publishing AG 2018
A. Konar and S. Saha, *Gesture Recognition*, Studies in Computational
Intelligence 724, DOI 10.1007/978-3-319-62212-5

Printed in the United States
By Bookmasters